Evaporative Cooling

The science of *beating the heat.*

by D. James Benton

Foreword

Evaporative cooling is vital for life and industry. Benefiting from this essential process is as natural as sweating. It's the most efficient means of rejecting unwanted heat. The economy of evaporative cooling has made it the first choice of manufacturing and power production. There is rarely time to adequately address this important topic in a course on heat transfer. Many articles and books present the mechanical and even economic aspects of evaporative cooling. This book presents the computational (i.e., mathematical) aspects. All of the software discussed in this book is available free on-line, including considerable source code.

All of the examples contained in this book,
(as well as a lot of free programs) are available at...

https://www.dudleybenton.altervista.org/software/index.html

Typical Spray

Typical Splash Bars

Table of Contents

page

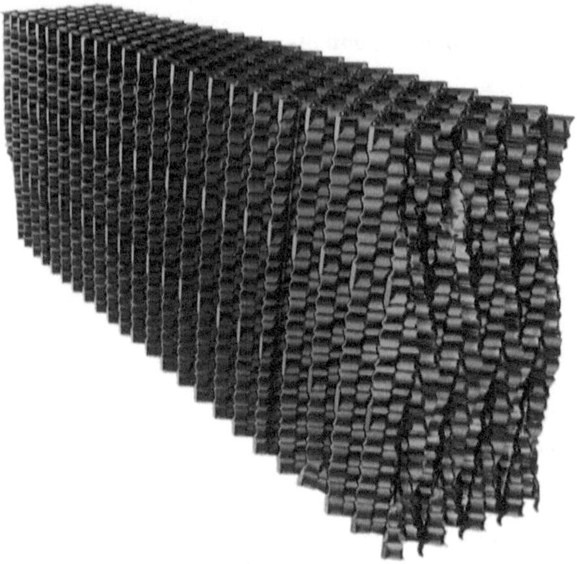

Typical Clogging-Resistant Film Fill

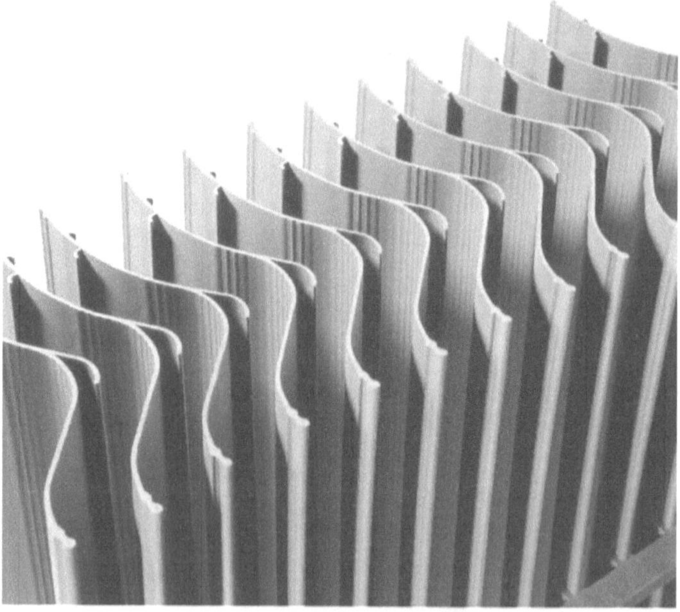

Typical Drift Eliminators

Chapter 1. Psychrometrics

Psychrometrics or psychrometry is the science of moist air, that is, the determination of physical and thermodynamic properties of air water vapor mixtures. The term comes from the Greek psuchron (ψυχρόν) meaning *cold* and metron (μέτρον) meaning *measurement*. This is where we begin the study of evaporative cooling.

The efficacy of evaporative cooling arises from the truly remarkable properties of the water molecule. This simple molecule appears in the solid, vapor, and liquid states, quite literally covering the surface of the Earth. Water has the highest latent heat of any naturally occurring substance and one of the highest specific heats. The only common molecule that comes close is ammonia, at a distant 60%. Next in line come the alcohols. Even so, ammonia and the alcohols don't exhibit all three states at nearly the same range of temperatures.

This remarkable behavior is due to the stress present in the unique structure of the water molecule. The hydrogen-oxygen bond is very strong, which draws them tightly together. The two hydrogen molecules repulse each other, which forces them apart. The end result is like a very tightly wound spring. This tug of war is illustrated in the following figure:

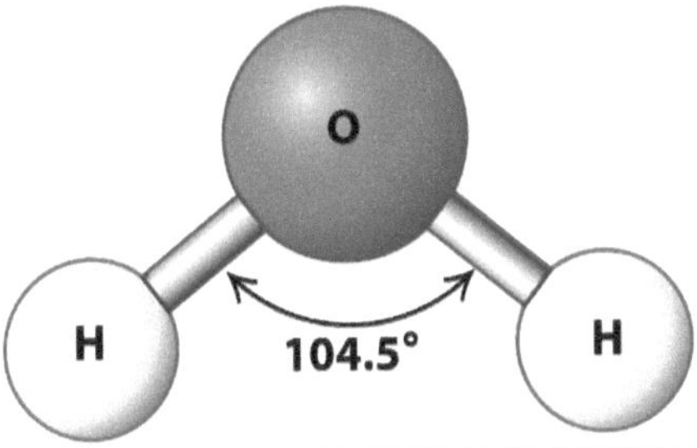

Mass Fraction of Water Vapor in Air by Temperature and Relative Humidity at 1 atm.

°F	0%	10%	20%	30%	40%	50%	60%	70%	80%	90%	100%
0	0.00%	0.01%	0.02%	0.02%	0.03%	0.04%	0.05%	0.06%	0.06%	0.07%	0.08%
10	0.00%	0.01%	0.03%	0.04%	0.05%	0.07%	0.08%	0.09%	0.11%	0.12%	0.13%
20	0.00%	0.02%	0.04%	0.06%	0.09%	0.11%	0.13%	0.15%	0.17%	0.19%	0.21%
30	0.00%	0.03%	0.07%	0.10%	0.14%	0.17%	0.21%	0.24%	0.28%	0.31%	0.34%
40	0.00%	0.05%	0.10%	0.16%	0.21%	0.26%	0.31%	0.36%	0.41%	0.47%	0.52%
50	0.00%	0.08%	0.15%	0.23%	0.30%	0.38%	0.46%	0.53%	0.61%	0.68%	0.76%
60	0.00%	0.11%	0.22%	0.33%	0.44%	0.55%	0.66%	0.77%	0.88%	0.99%	1.10%
70	0.00%	0.15%	0.31%	0.46%	0.62%	0.78%	0.93%	1.09%	1.24%	1.40%	1.56%
80	0.00%	0.22%	0.43%	0.65%	0.87%	1.09%	1.30%	1.52%	1.74%	1.96%	2.19%
90	0.00%	0.30%	0.60%	0.90%	1.20%	1.50%	1.80%	2.11%	2.41%	2.72%	3.03%
100	0.00%	0.41%	0.81%	1.22%	1.63%	2.05%	2.46%	2.88%	3.30%	3.72%	4.14%
110	0.00%	0.54%	1.09%	1.64%	2.20%	2.76%	3.32%	3.89%	4.46%	5.04%	5.61%
120	0.00%	0.72%	1.45%	2.19%	2.94%	3.69%	4.44%	5.21%	5.98%	6.76%	7.54%

Because the latent heat of water is so large, evaporating even a little of it into the air caries with it a surprising amount of heat. The first table shows the mass fraction of water vapor in air for a range of temperature and relative humidity, which varies from 0 to 7.54%: The second table shows the energy fraction, that is, the fraction of the total energy (air plus water vapor) that is present in the water vapor alone:

Energy Fraction of Water Vapor in Air by Temperature and Relative Humidity at 1 atm.											
°F	0%	10%	20%	30%	40%	50%	60%	70%	80%	90%	100%
0	1.1%	5.0%	9.6%	13.7%	17.5%	21.0%	24.1%	27.1%	29.8%	32.3%	34.6%
10	1.3%	5.5%	10.4%	14.9%	18.9%	22.6%	25.9%	29.0%	31.8%	34.5%	36.9%
20	0.9%	4.6%	8.7%	12.6%	16.1%	19.3%	22.3%	25.1%	27.7%	30.2%	32.4%
30	1.0%	4.9%	9.3%	13.3%	17.0%	20.4%	23.6%	26.5%	29.2%	31.7%	34.0%
40	1.3%	5.5%	10.4%	14.9%	18.9%	22.6%	26.0%	29.0%	31.9%	34.5%	37.0%
50	1.7%	6.4%	12.0%	17.1%	21.5%	25.6%	29.2%	32.5%	35.6%	38.3%	40.9%
60	2.3%	7.6%	14.2%	19.9%	24.9%	29.3%	33.3%	36.8%	40.0%	42.9%	45.6%
70	3.1%	9.1%	16.8%	23.3%	28.8%	33.7%	37.9%	41.7%	45.0%	48.0%	50.7%
80	4.3%	11.0%	19.9%	27.2%	33.3%	38.5%	43.0%	46.9%	50.3%	53.4%	56.1%
90	5.8%	13.2%	23.4%	31.5%	38.2%	43.7%	48.3%	52.3%	55.7%	58.7%	61.4%
100	7.8%	15.8%	27.4%	36.3%	43.3%	49.0%	53.7%	57.7%	61.1%	64.0%	66.6%
110	10.3%	18.7%	31.7%	41.3%	48.6%	54.4%	59.1%	63.0%	66.2%	69.0%	71.4%
120	13.2%	22.0%	36.4%	46.4%	53.9%	59.7%	64.3%	68.0%	71.1%	73.7%	75.9%

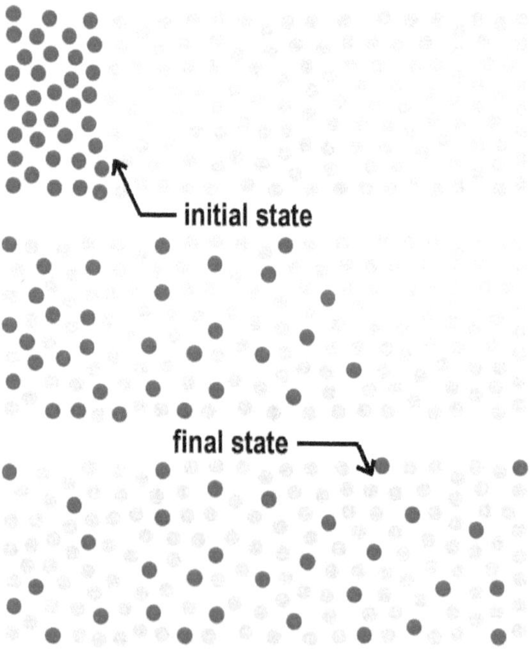

The states highlighted in blue have more energy in the water vapor than the air. Water vapor accounts for only 7.54% of the mass in the lower right corner, but 75.9% of the energy. Because moist air can contain this remarkable amount of energy, it can receive it through an evaporative process. In such a process, the water molecules must diffuse into the air. While the air-water diffusion coefficient isn't among the largest, it is large enough to readily facilitate natural and industrial processes. Diffusion is easily visualized by the spreading of dye in water. What happens on a molecular level is illustrated in the figure, where the blue dots represent water molecules and the yellow dots represent air molecules:

The properties of moist air are defined by the American Society of Heating, Refrigeration, and Air Conditioning Engineers (ASHRAE) in their Handbook of Fundamentals, which is published periodically. This is the de-facto worldwide standard. The reference point for these properties is as follows: 1) The enthalpy of dry air is zero at 0°F; and 2) The enthalpy of water vapor is zero at the triple point (i.e., 32.018°F). This difference in reference points for the dry air and water vapor is not a problem, as long as you are dealing with moist air at near atmospheric conditions and use this same formulation throughout.

This is not always the case, as we shall see in an application later on in the book. The problem comes when you mix moist air with some other gases or when the temperature is near or above 212°F/100°C. In such cases, a different formulation is necessary. The ASHRAE formulation is one of convenience. It's simply easier to handle calculations for applications that fit into this category. The ASHRAE formulation is not any less accurate for making this selection of reference conditions.

Dry-Bulb, Wet-Bulb, and Dew Point

The terms *dry* and *wet* are simply that and the term *bulb* arose in the era of mercury-in-glass thermometers. The dry-bulb is the ordinary air temperature that is cited in the evening weather report. The wet-bulb is found by attaching a wick (i.e., a little sock) on the end of a thermometer, keeping it wet, and gently blowing air across it. The *wet-bulb depression* is the difference between the dry-bulb and wet-bulb temperatures.

If the air is very dry, the water will quickly evaporate from the wick so that the temperature of the thermometer will drop. If the air is already very humid, the water will slowly evaporate, it at all, resulting in no drop in temperature. The recommended air velocity for wet-bulb measurement is 2.5 m/s (8.2 ft/sec). Such a device should be shielded from thermal radiation.

The purpose of wet-bulb measurement is to approximate the adiabatic (i.e., no heat transfer) saturation process. This means brining the air from it's ambient state to one of being saturated with water vapor without transferring any heat. In so much as the wet-bulb approximates the adiabatic saturation temperature, it is a measurement of the enthalpy. The temperature and enthalpy are independent properties; therefore, measurement of these two uniquely determines the state of the air, including the mass or mole fraction of the water vapor.

It is commonly observed that a hot dry day is not as uncomfortable as a hot humid one. For example, 110°F in Phoenix may be considered more comfortable than 90°F in New Orleans. This is because skin is porous and contains sweat glands, so that humans' perception of temperature is closer to the wet-bulb than the dry-bulb. It's the enthalpy, not the temperature, of air that determines how readily it will receive the heat your body must reject. A typical box-style psychrometer is shown in the next two figures:

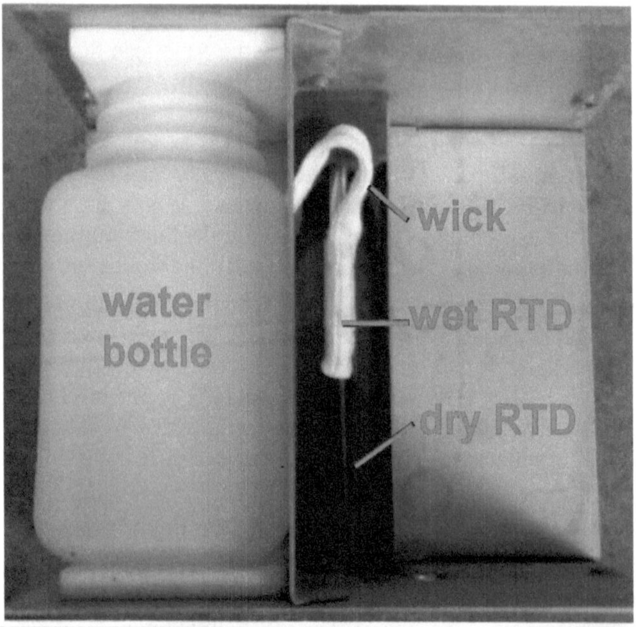

Dew point is found by cooling the air to the point where condensation begins. For completely dry air, this will never happen, because there is no water vapor to condense. Dew point is simply a measure of the water vapor content and is not a measure of the enthalpy. Dew point is often included in the weather report, because it can be conveniently measured and the instruments don't require nearly the maintenance of a wet-bulber. Dew point is measured with a *chilled mirror hygrometer*, which is illustrated conceptually in this next figure:

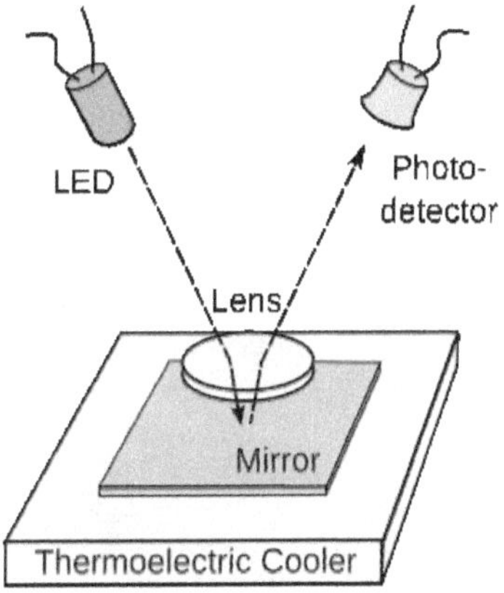

The actual device looks like the following:

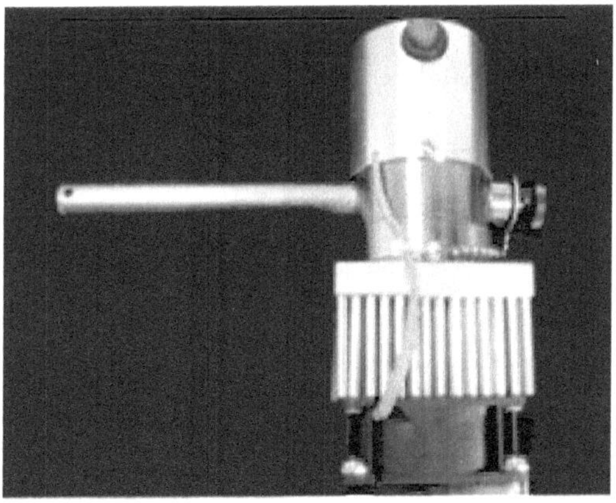

There is a little fan at the bottom (black barrel shape). On top of this is a finned aluminum heat sink. The mirror and photo detector are in the cylinder at the top. The thermocouple or RTD (resistance thermal detector) is also inside.

Under most conditions the dry-bulb will be higher than the wet-bulb, which will be higher than the dew point. The only time the three are the same is at saturation (i.e., 100% relative humidity). There are some odd cases, for instance, in a fog, the dew point is higher than the dry-bulb, as you would need to slightly heat the air to reach the point where condensation *first* begins.

Specific vs. Relative Humidity

The *specific humidity* is expressed as grains (7000 grains = 1 pound) per pound of dry air and is an antiquated term, though some instruments still report humidity in this form. The *humidity ratio* is simply the pounds (or kg) of water vapor per pound (or kg) of dry air. This is numerically equal to the specific humidity in grains divided by 7000. Humidity ratio is usually given the symbol "***W***" or "ω".

The most familiar term is *relative humidity.* This is the ratio of the actual water vapor content to the maximum (or saturation) value and is most often reported in percent. Completely dry air (devoid of any water vapor) has a relative humidity of 0%. When it's raining or foggy, the relative humidity is 100%. The amount of water vapor that is contained in air at saturation varies across orders of magnitude for the range of temperatures commonly present on the face of the Earth.

Chapter 2. Thermodynamic Properties of Moist Air

Precise determination of the thermodynamic properties of moist air in the West began with the work of Goff & Gratch, published in a series of papers between 1943 and 1949.[1,2] This work was continued by Hyland & Wexler between the years 1978 and 1983.[3,4,5] Nelson and Sauer further refined with their research between 1999 and 2001.[6] Most recently, Hermann, Kretzschmar, and Gatley have presented a more complete formulation coupled with the latest properties of steam.[7,8]

The basic formulation has been the same since the work of Goff & Gratch. The significant peculiarity of this approach is that the properties are all on a per pound of dry air basis, rather than on a per total (air plus water vapor basis). This facilitates some calculations, which was of greater concern in the 1940s than it is now. It doesn't matter, as long as you are consistent. For the most part these properties are used for atmospheric processes, which is not a problem. As mentioned in the previous chapter and in a subsequent example, it does matter at elevated temperatures (near or above 212°F/100°C).

Saturation Pressure and the Enhancement Factor

It is tempting to simply use the saturation pressure of steam along with Dalton's Law of Partial Pressures[9] to obtain values for the water vapor content in air at saturation, but this isn't accurate. The saturation pressure of steam is for

---

[1] Goff, J. A. and Gratch, S., "Thermodynamic Properties of Moist Air," ***Heating, Piping & Air Conditioning***, pp. 334-348, 1945.

[2] Goff, J. A. and Gratch, S., "Low-Pressure Properties of Water from -160 to 212 F," ASHVE Trans., pp. 95-122, 1946.

[3] Hyland, R. W., Wexler, A., and Stewart, R., "Thermodynamic Properties of Dry Air, Moist Air and Water and SI Psychrometric Charts," ASHRAE RP-216 and RP-25, 1983.

[4] Hyland, R. W. and Wexler, A., "Formulations for the Thermodynamic Properties of the Saturated Phases of H2O from 173.15 K to 473.15 K," ASHRAE Trans., Vol. 89, pp. 500-519, 1983.

[5] Hyland, R. W. and Wexler, A., "Formulations for the Thermodynamic Properties of Dry Air from 173.15 K to 473.15 K, and of Saturated Moist Air from 173.15 K to 372.15 K, at Pressures to 5 MPa," ASHRAE Trans., Vol. 89, pp. 520-535, 1983.

[6] Nelson, H. F. and Sauer, H. J., "Formulation of High-Temperature Properties for Moist Air," ***HVAC&R Research*** Vol. 8, pp. 311-334, 2002.

[7] Herrmann, S., Kretzschmar, H.-J., and Gatley, D. P., "Thermodynamic Properties of Real Moist Air, Dry Air, Steam, Water, and Ice," ***HVAC&R Research***, 2009.

[8] Herrmann, S., Kretzschmar, H.-J., and Gatley, D. P., "Thermodynamic Properties of Real Moist Air, Dry Air, Steam, Water, and Ice - Final Report," ASHRAE RP-1485, 2009.

[9] Dalton's Law of Partial Pressures states that, in a mixture of non-reacting gases, the total pressure is equal to the sum of the partial pressures exerted by each of the constituents and these individual contributions to the whole are each proportional to the mole fraction of that component.

$H2O$ in the vapor state in equilibrium with $H2O$ in the liquid state. This isn't the same as $H2O$ in the vapor state in equilibrium with air in the gaseous state. An *enhancement factor*, *f*, is introduced to account for this difference. The enhancement factor is equal to the partial pressure of water vapor that should produce the observed content divided by the saturation pressure of steam at that same temperature. The values of *f* are close to unity and the symbol is appropriate, as this is simply a *fudge factor.*

Herrmann, Kretzschmar, and Gatley, present a very complicated equation in Section 3.4.2.1 of their report for ***ln(f)*** in terms of second and third virial coefficients[10], explaining that this arises from Henry's Law.[11] While this is interesting and may facilitate the calculation of *f* at elevated pressures without necessitating experiments, it is immaterial. All that is needed is to measure and tabulate the water content of air. An explanation as to why it is what it is, is not essential. This is the approach that Goff & Gratch and Hyland & Wexler took. The following table of f can be found in any edition of the *ASHRAE Handbook of Fundamentals.*

Enhancement Factor, f

	Pressure [in.Hg]					
T°F	10	15	20	25	30	32
0	1.0016	1.0025	1.0033	1.0040	1.0047	1.0051
20	1.0016	1.0024	1.0032	1.0039	1.0045	1.0048
40	1.0018	1.0025	1.0032	1.0038	1.0044	1.0047
60	1.0020	1.0026	1.0033	1.0039	1.0044	1.0047
80	1.0023	1.0029	1.0036	1.0041	1.0046	1.0049
100	1.0027	1.0033	1.0040	1.0045	1.0050	1.0053
120	1.0031	1.0037	1.0044	1.0050	1.0055	1.0057
140		1.0041	1.0048	1.0054	1.0059	1.0063

The variation of *f* with temperature at 1 atm. is shown in the following. For the purpose of calculations, the humidity ratio, ***W***, is needed. This can be calculated from *f*, ***Psat***, and the molecular weights by Equation 2.1:

$$W = \left(\frac{MW_{H2O}}{MW_{AIR}}\right)\left(\frac{fP_{SAT}}{P_{BARO} - fP_{SAT}}\right) \tag{2.1}$$

The molecular weight of water is 18.01528 and of air is 28.9645. ***Pbaro*** is the barometric pressure. ***Psat*** is the saturation pressure of steam in the same units as the barometric pressure. The denominator in Equation 2.1 becomes zero

---

[10] The virial expansion of the equation of state was first proposed by Kamerlingh Onnes in 1901. It forms the basis for many developments in thermodynamics related to the properties of fluids. It is... $Z=PV/RT=1+B\rho+C\rho^2+\ldots$

[11] Henry's Law states that, at a constant temperature, the amount of a gas that will dissolve in a liquid is directly proportional to the partial pressure of that gas in equilibrium with that liquid. William Henry 1803.

when ***fPsat=Pbaro***, which is why this formulation can't be used at elevated temperatures.

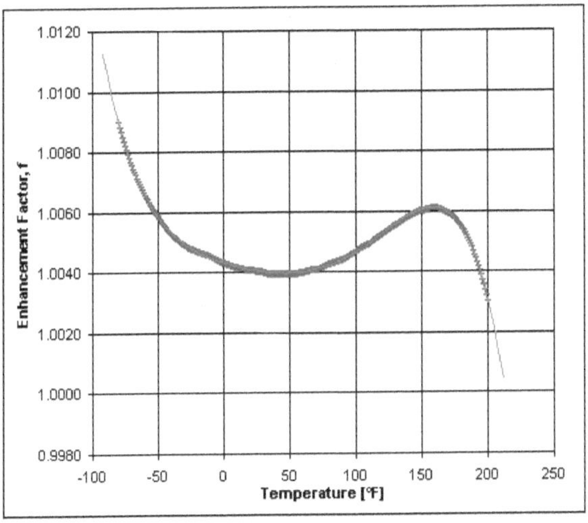

The variation of ***Psat*** and ***W*** with temperature at 1 atmosphere barometric pressure is shown in this next figure:

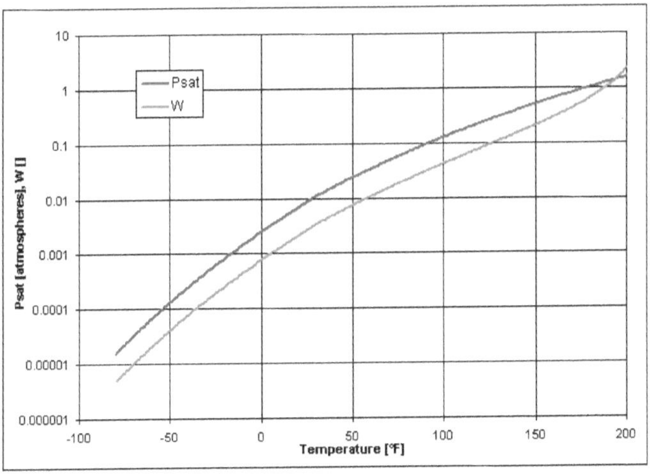

<u>Enthalpy and Entropy</u>

The enthalpy of moist air is also calculated on a per unit mass of dry air basis. Over the range of interest (-80°F to 212°F/-62°C to 100°C), the specific heat, ***Cp***, of air varies so little that 0.24 BTU/lbm/°F (1 kJ/kg/°C) is an adequate representation. The enthalpy of water vapor varies linearly over this range so that the following equation is adequate:

$$h_G = 1061 + 0.444T \tag{2.2}$$

Temperature is in degrees Fahrenheit and enthalpy is in BTU/lbm. Conversion to degrees Celsius and kJ/kg is trivial. It is very important to stress here that the appropriate enthalpy of water vapor is $\boldsymbol{h_G}$ and NOT $\boldsymbol{h_{FG}}$ (that is, the enthalpy of the saturated vapor, not the latent heat of vaporization or the difference between the vapor and liquid enthalpies). Although many will insist the latter is correct-this statement even appears in print-it is not true. Consider the case of steam at the critical point flowing into the room where you now sit. At the critical point $\boldsymbol{h_{FG}=0}$. You should leave immediately. You'll be thoroughly cooked long before the temperature reaches 705°F, which it most assuredly will. Energy is entering the room, but $\boldsymbol{h_{FG}=0}$.

The full equation for enthalpy is:

$$h = h_A + Wh_G = 0.24\,T + W(1061 + 0.444\,T) \tag{2.3}$$

The entropy is a little more complicated, because of the partial pressures:

$$s_A = 0.24\ln\left(\frac{T + 469.67}{469.67}\right) \tag{2.4}$$

$$s_G = 2.29688 - 0.00369268T + 0.0000055T^2 \tag{2.5}$$

$$s = s_A + Ws_G - R\ln\left(\frac{P}{14.696}\right) \tag{2.6}$$

A few values are listed in the following table:

T	Ws	Ps	f	Ha	Hg	Hs	Sa	Sg	Ss
-80	0.000004948	0.000236	1.009015	-19.221	1212.6	-19.215	-0.046	-11.7	-0.046
-60	0.00002121	0.001013	1.006593	-14.414	1037.2	-14.392	-0.034	0.7	-0.034
-40	0.00007929	0.003793	1.005314	-9.609	1046.8	-9.526	-0.022	2.0	-0.022
-20	0.0002632	0.012595	1.004719	-4.804	1052.4	-4.527	-0.011	2.1	-0.010
0	0.0007875	0.037671	1.004350	0.000	1060.3	0.835	0.000	2.2	0.002
20	0.0021531	0.102798	1.004101	4.804	1069.6	7.107	0.010	2.2	0.015
40	0.005216	0.24784	1.003959	9.609	1078.2	15.233	0.020	2.159	0.031
60	0.011087	0.52193	1.004025	14.415	1087.0	26.467	0.029	2.094	0.053
80	0.022340	1.03302	1.004285	19.222	1095.7	43.701	0.039	2.035	0.084
100	0.043219	1.93492	1.004706	24.031	1104.4	71.761	0.047	1.982	0.133
120	0.081560	3.45052	1.005246	28.842	1112.9	119.612	0.056	1.933	0.213
140	0.153538	5.88945	1.005830	33.656	1121.3	205.824	0.064	1.889	0.354
160	0.29945	9.6648	1.006120	38.474	1129.6	376.737	0.072	1.848	0.625
180	0.65911	15.3097	1.005510	43.295	1137.7	793.166	0.079	1.811	1.273
200	2.30454	23.4906	1.003033	48.121	1145.6	2688.205	0.087	1.776	4.180

In this table *a* denotes air, *g* denotes gas (or vapor), and *s* denotes saturation. Functions to calculate the properties of moist are may be found in Appendix B and are included in the on-line archive.

Chapter 3. Merkel's Equation

In 1925 Merkel proposed a theory relating the evaporation and sensible heat transfer occurring in a direct contact process such as cooling of water or humidification of air, to an air enthalpy difference.[12] Such a representation was suited (but not limited to) various types of cooling towers. The derivation was based on counterflow contact of water and air. In fact, there were six basic assumptions that were introduced at various points in the development to simplify the mathematics. This chapter is taken largely from a paper the author published with Al Feltzin in 1991, which I suggest for further reading.[13]

The model on which Merkel's theory was developed consists of a water droplet at temperature, ***T***, surrounded by a thin air film (interface). He assumed that the air film is saturated and therefore is also at temperature T_A. Thus the film has humidity ratio, W_F, and enthalpy, h_F. Surrounding the air film is the bulk air mass at some lower temperature $T_A<T_F$, and humidity ratio, $W_A<W_F$, and an enthalpy $h_A<h_F$.

If ***a*** is the interfacial surface (ft²/ft³) and ***V*** is the contacting volume (ft³) then the interfacial surface area, ***S=aV*** (ft) and the differential surface of the model droplet interfacial film is ***dS=adV*** (surface element). The downward flux of water is given the symbol, ***L*** (lbs/hr/ft²) and the upward flux of air, ***G*** (lbs/hr/ft²). The two fluxes are opposing, that is, *counterflow*. The following figure illustrates this process and the variables:

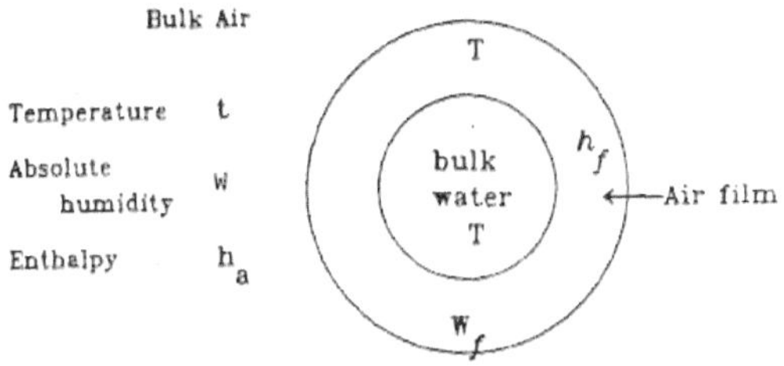

Heat is transferred from the water droplet to the bulk air through the interface by two means, sensible heat transfer (convection, due to a difference in temperature) and latent heat of evaporation (mass transfer by diffusion, due to a difference in concentration). Merkel assumed that the interface offers no resistance to heat transfer from the water droplet to the bulk air by either of these mechanisms. The sensible heat transfer rate by convection is given by:

$$dq_S = K_C(T_F - T_A)adV \tag{3.1}$$

---

[12] Merkel, F. Verdunstungskuehlung, V.D.I. Forschungsarbeiten, No. 275, Berlin, 1925.
[13] Feltzin, A. E, and Benton, D. J., "A More Nearly Exact Representation of Cooling Tower Theory," CTI Technical Paper Number TP91-02.

where K_C is the convective heat transfer coefficient, BTU/hr/ft²/F. The mass transfer rate is given by:

$$dL = K_M\left(P_F - P_A\right)adV \tag{3.2}$$

where K_M is a diffusional mass transfer coefficient. P_F and P_A are the partial pressures of water vapor in the interfacial film at temperature, T_F, and bulk air at temperature, T_A, respectively. Next, Merkel assumed that the partial pressure of water vapor is proportional to humidity, that is:

$$\begin{aligned} P_F &\propto W_F \\ P_A &\propto W_A \end{aligned} \tag{3.3}$$

which can be substituted into Equation 3.2 to obtain:

$$dL = K_M\left(W_F - W_A\right)adV \tag{3.4}$$

The evaporative (latent) heat transfer rate due to diffusional mass transfer is given by:

$$dq_L = \lambda dL = \lambda K_M\left(W_F - W_A\right)adV \tag{3.5}$$

where λ is the latent heat of vaporization. The total transfer rate is then given by:

$$dq_{total} = \left[K_C\left(T_F - T_A\right) + \lambda K_M\left(W_F - W_A\right)adV\right] \tag{3.6}$$

At this point in the derivation, the concept of humid heat, C_S, (the heat capacity of an air-water vapor mixture) is usually introduced. By addition and subtraction of a term $C_S(T_W - T_A)$ to the right hand side of Equation 3.6 and algebraic manipulation (see Merkel Appendix A), we arrive at the following:

$$dq_{total} = K_M\left\{\left(C_S T_F + \lambda W_F\right) - \left(C_S T_A + \lambda W_A\right) + C_S\left(T_F - T_A\right)\left[\frac{K_C - 1}{C_S K_M}\right]\right\}adV \tag{3.7}$$

Merkel also assumed that the Lewis Number of unity, that is:

$$L_E = \frac{K_C}{C_P K_M} = 1 \tag{3.8}$$

This assumption causes the last term in Equation 3.7 to vanish. The terms $C_S T_F + \lambda W_F$ and $C_S T_A + \lambda W_A$ are close to, but not exactly, equal to Equation 2.3, reducing the heat transfer to:

$$dq = K\left(h_S - h_A\right)\, adV \tag{3.9}$$

The subscript on K has been dropped, as there is now only one. The subscript F (film) has been replaced with S (saturation), in keeping with the previous assumption. Conservation of energy requires that:

$$dq = d\left(LC_{PW}T_W\right) = d\left(Gh_A\right) \tag{3.10}$$

where C_{PW} is the constant pressure specific heat of water in the liquid state. These two equations can be combined to form:

$$dq = C_{PW}\left(LdT_W - T_W dL\right) = Gdh_A = K\left(h_F - h_A\right)adV \qquad (3.11)$$

Dividing by h_F-h_A and integrating yields:

$$KaV = C_{PW}\left\{\int_{T_{OUT}}^{T_{IN}} \frac{LdT}{\left(h_F - h_A\right)} + \int_{T_{OUT}}^{T_{IN}} \frac{TdL}{\left(h_F - h_A\right)}\right\} \qquad (3.12)$$

$$KaV = G\int_{h_{ain}}^{h_{aout}} \frac{dh_A}{\left(h_F - h_A\right)} \qquad (3.13)$$

Merkel went on to assume that the portion of the water evaporated was insignificant to the whole, or $dL=0$. After making this assumption, Equations 3.12 and 3.13 become:

$$\frac{KaV}{L} = C_{PW}\int_{T_{OUT}}^{T_{IN}} \frac{dT}{h_F - h_A} = \frac{G}{L}\int_{h_{ain}}^{h_{aout}} \frac{dh}{\left(h_F - h_A\right)} \qquad (3.14)$$

The third term in Equation 3.14 is of no further interest at this point, but will be utilized later. This leaves the classic form of Merkel's Equation:

$$\frac{KaV}{L} = C_{PW}\int_{T_{OUT}}^{T_{IN}} \frac{dT}{h_F - h_A} \qquad (3.15)$$

This is a nonlinear equation with no analytical solution and must be solved numerically. Merkel chose to use the 4-point Chebyshev method for its simplicity. In spite of these assumptions and simplifications, Merkel's Equation served the cooling tower industry well for decades.

Over the years, several approaches have been devised in an attempt to compensate for several of the above assumptions and approximations. Mickley[14] introduced temperature and humidity gradient, heat and mass transfer coefficients from water to interfacial film and from film to air. Baker & Mart[15] developed a hot water correction factor that reduced the scatter in test data. Temperature correction factors have been applied to demand curves in some cases and fill (characteristic) curves in others, this has resulted in a confusing situation at best. Effects of air temperature, barometric pressure, and salinity on fill characteristic KaV/L have been discussed by LeFevre.[16] Baker & Shryock discuss the method of integration at some length.[17]

---

[14] Mickley, H. S., *Chemical Engineering Progress*, Vol. 45, p. 739, 1949.
[15] Baker, R. and Mart, L T., *Refrigeration Engineering*, p. 965, 1952.
[16] LeFevre, M., CTI Paper 58, New Orleans, Louisiana, 1985.
[17] Baker & Shryock, *Journal of Heat Transfer*, Appendix B and Table lB, April, 1961.

Chapter 4. Counterflow Demand Curves

The Cooling Tower Institute[18] using Equation 3.15 and the 4-point Chebyshev, expanded on earlier KaV/L vs. L/G demand curves such as those utilized by Foster Wheeler Corporation[19] and J.F. Pritchard Company.[20] The CTI curves had the advantages of being computer generated and computer drawn, and made what had been very limited published data much more widely available and over a much broader selection of ranges and approaches. These curves served the industry for over 20 years. The following is typical:

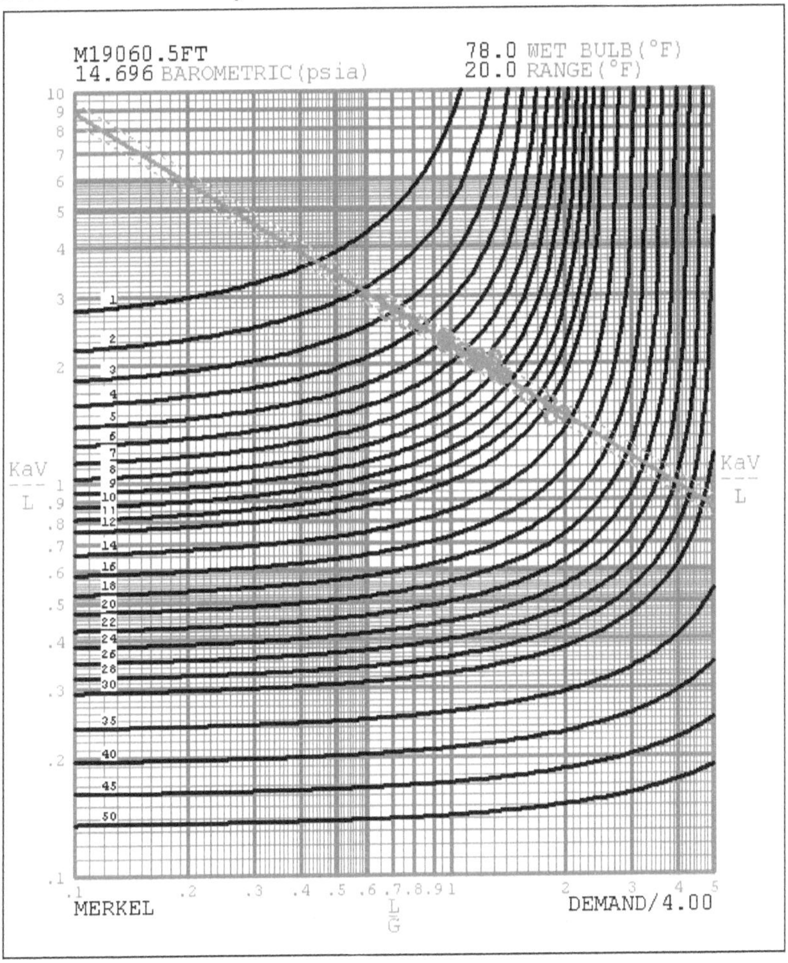

---

[18] *CTI Blue Book of Counterflow Demand Cooling Curves*, CTI, Houston, Texas, 1967.
[19] *Cooling Tower Performance: Bulletin CT432*, Foster Wheeler Corporation, 1943.
[20] *Counterflow Cooling Tower Performance*, F. Pritchard Company of California, 1957.

These *demand* curves were used to obtain graphical solutions of cooling tower performance, which was a tedious process. The loose-leaf notebook filled with such curves was an expensive item and hard to come by. The same curves can now easily be generated with an Excel® spreadsheet, as illustrated below. This spreadsheet is included in the on-line archive. A program to generate the professional grade curves is available free on-line from the author's web site.

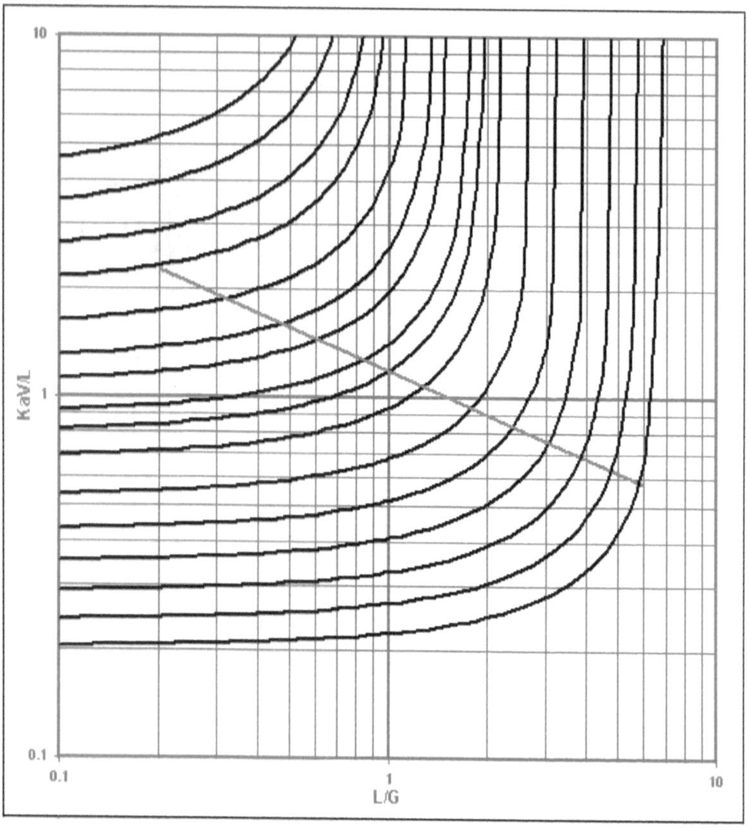

In these figures, the black curves are constant *approach* (cold water temperature minus ambient wet-bulb). The horizontal axis is the ratio of the water to air mass flows. The downward-sloping red line in the previous figure is the *supply* line, or the dimensionless mass transfer capacity provided by the packing material. Generating performance curves for a cooling tower using this graphical method was beyond tedious. This can also readily be done with an Excel® spreadsheet as illustrated in this next figure. This is on another tab in the same spreadsheet that comes with the archive.

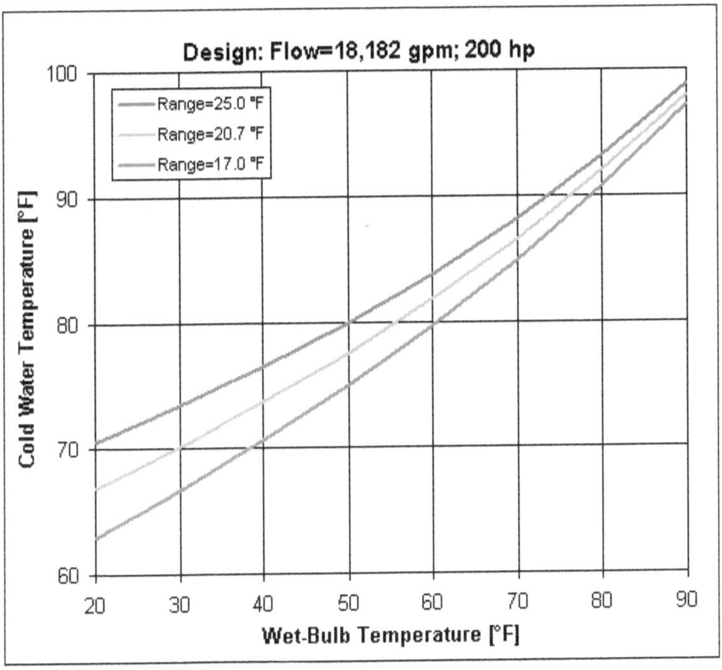

The *range* is the hot (entering) water temperature minus the cold (leaving) water temperature. Only the wet-bulb is needed here, because the airflow is provided by fans, making this a mechanical-draft cooling tower. The same formula and calculation can be used for natural-draft (buoyancy-induced airflow), as will be illustrated subsequently.

So far we have only considered the counterflow arrangement. There are operational and economic reasons for designing cooling towers with a crossflow arrangement. Solving the crossflow mass transfer equation requires a different approach, one that was quite a challenge in the 1960s with limited computational options.

Chapter 5. Crossflow Demand Curves

The counterflow problem can be expressed as a single integral equation. Although this is nonlinear, the solution is simple. This is not true for a crossflow problem. Not only must the integration be carried out over two dimensions, there are more computational problems that arise, none the least of which is a very strong tendency to over-shoot, which leads to instability and erroneous solutions. The crossflow calculations are also more complex to visualize. The following spreadsheet is arranged to help visualize the calculations:

Crossflow Cooling Tower Demand Calculations

user inputs
50 Twb
120 Thw
2.00 L/G
1.38 KaY
calculations
83.2 Tcw
36.8 range
33.2 approach

						Tw				
Ta						120.0	120.0	120.0	120.0	120.0
50	102	113	117	119	119	92.6	107.7	114.5	117.5	118.9
50	78	97	108	113	117	81.7	95.2	104.3	110.6	114.7
50	70	86	97	105	111	74.7	87.1	96.2	103.2	108.5
50	65	79	89	98	104	69.8	81.1	89.9	96.9	102.6
50	62	74	83	91	98	66.0	76.4	84.7	91.7	97.4
						Hw				
Ha						119.7	119.7	119.7	119.7	119.7
20.3	75.2	99.7	110.7	115.7	117.9	59.6	87.2	103.7	112.2	116.3
20.3	42.0	66.9	87.2	101.0	109.4	45.6	63.7	80.1	93.9	104.2
20.3	34.3	50.5	66.8	81.8	94.2	38.4	52.1	65.3	77.7	88.9
20.3	30.3	42.3	55.0	67.5	79.3	33.9	44.9	55.8	66.4	76.6
20.3	27.8	37.2	47.5	57.9	68.2	30.8	39.9	49.1	58.3	67.3

Each box represents one computational cell in a 5x5 grid. The air flows in from the left and the water flows down from the top. There are several variables associated with each cell. These are arranged in groups and distinguished by colors.

The light blue cells under the heading "Ta" are the air temperature (wet-bulb). This group is 5x6, because for each cell there is an inlet and exit value, requiring one extra column. The light magenta cells under the heading "Tw" are the water temperature. This group is 6x5, because for each cell there is an inlet and exit value, requiring one extra row. The green and yellow cells contain the enthalpies for the air and water respectively. Hw is the enthalpy of saturated air at the water temperature, not the enthalpy of the water.

The difference between the air and water enthalpies in each cell is the potential driving the mass transfer in that cell. Each cell contains 1/25th of the packing material and surface area for the exchange. The product of the potential, the transfer coefficient, and surface area is equal to the heat transfer, Q. The exit water temperatures in the magenta section and exit air enthalpies in the green section are calculated from Q for each cell. The enthalpies in the yellow section corresponding to the exit water conditions are calculated from the water temperatures in the magenta section.

There are two tabs in the spreadsheet. One contains a 5x5 cell calculation and the other a 14x14. You can compare these two in order to get an idea of how many cells it takes to obtain a sufficiently accurate representation of this problem. The spreadsheet is included in the on-line archive and can be used to get a better understanding of the computer modeling process. The same free program that generated the counterflow demand curves will also generate crossflow curves, as illustrated in the next figure:

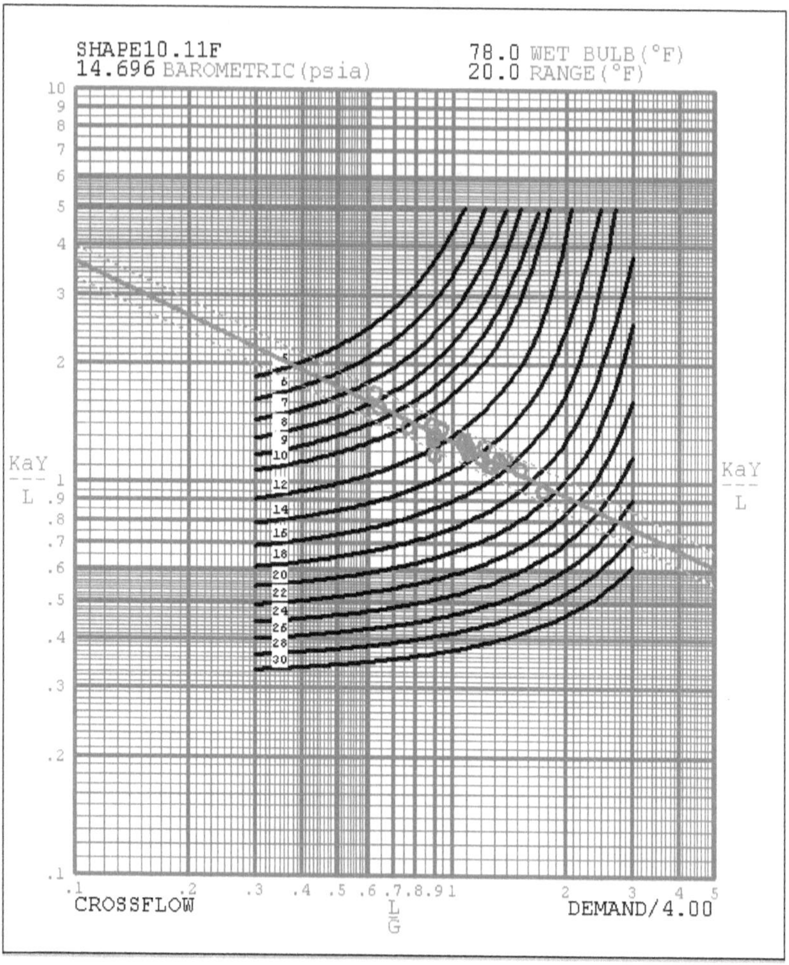

Chapter 6. A Closer Look at Crossflow

Modeling the heat and mass transfer process in an evaporative cooling tower is an exercise in applied mathematics wherein a system of differential equations is solved. There are many methods for solving such equations. The content of this chapter is a condensation of several publications that are available free on-line and which I suggest for further reading.[21,22,23,24] We begin with a schematic of the computational grid, in this case for a crossflow arrangement:

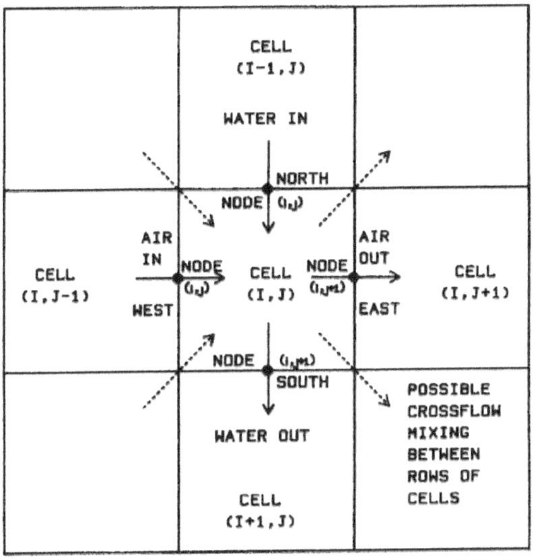

As in the multi-colored spreadsheet from the previous chapter, water enters the cells from the top and exits at the bottom. The air enters the cells from the left and exits to the right. At this point we will ignore redistribution of the airflow through the fill. We will also assume that both the air and water flux (mass flow per unit area) is uniform throughout the fill.

As the inlet conditions are required for the calculations and the exit conditions are the result, we simply step through the fill horizontally, shift down

---

[21] D. J. Benton, "A Numerical Simulation of Heat Transfer in Evaporative Cooling Towers," TVA Report No. WR28-1-900-110, September, 1983.
[22] D. J. Benton, "Development of the Finite-Integral Method," TVA Report No. WR2B-2-900-148, December, 1984.
[23] D. J. Benton, "Computer Simulation of Hybrid Fill in Crossflow Mechanical-Induced-Draft Cooling Towers," Proceedings of the ASME Winter Annual Meeting, New Orleans, Louisiana, December 9-14, 1984.
[24] D. J. Benton and W. R. Waldrop, "Computer Simulation of Transport Phenomena in Evaporative Cooling Towers," ASME Journal of Engineering for Gas Turbines and Power, Vol. 110(2), pp. 190-196, April, 1988.

one row of cells, step through horizontally again, and continue this process to the end. The simplest method for solving the differential equation is the 4th order Runge-Kutta method. Implementation is easy, requiring only a few lines of code:

```
void Cell(double X,double Y,double Ha,double*dHa,double
    Tw,double*dTw)
  {
  double Hw,Q;
  Hw=fHtwb(Pbaro,Tw);
  Q=KaY*(Hw-Ha);
  dHa[0]=Q*LG;
  dTw[0]=-Q;
  }
```

The dependent variables, X and Y, are the position within the fill and aren't used in this case. There are two independent variables: 1) the enthalpy of the air and 2) the temperature of the water. These are passed in Ha and Tw, respectively. The differentials, dHa/dX and dTw/dY are returned in dHa and dTw, respectively. The heat transfer, Q, is out of the water and into the air. The ratio of the water to airflow rates, L/G, is applied to the air side.

As Merkel's approximations are assumed, the potential for driving the mass transfer is proportional to the enthalpy difference. The air enthalpies are initialized left face and water temperatures on the top face. The exiting water temperatures are averaged, yielding the cold water temperature, range, and approach. The following output is typical for a 9x9 grid:

```
grid: 9x9, KaY/L=1.38, L/G=2
************ Water Temps ************ avg. exit 89.0
120.0 120.0 120.0 120.0 120.0 120.0 120.0 120.0 120.0
109.2 111.6 113.4 114.9 116.0 116.8 117.5 118.0 118.5
101.3 104.9 107.8 110.1 111.9 113.4 114.6 115.6 116.4
 95.0  99.3 102.8 105.7 108.1 110.0 111.7 113.0 114.2
 89.9  94.6  98.5 101.8 104.5 106.8 108.7 110.4 111.8
 85.6  90.6  94.7  98.2 101.2 103.7 105.9 107.8 109.4
 82.0  87.1  91.4  95.0  98.1 100.8 103.2 105.2 107.0
 78.9  84.0  88.4  92.1  95.3  98.2 100.6 102.8 104.7
 76.2  81.3  85.7  89.4  92.7  95.6  98.2 100.5 102.5
 73.8  78.8  83.2  87.0  90.3  93.3  95.9  98.3 100.4
**************** Air Temps **************** avg. exit 105.4
 50.0  78.2  91.9 100.0 105.3 109.0 111.6 113.6 115.1 116.2
 50.0  72.5  85.2  93.5  99.4 103.7 107.0 109.5 111.6 113.2
 50.0  68.5  80.2  88.4  94.5  99.1 102.8 105.7 108.1 110.1
 50.0  65.6  76.3  84.2  90.3  95.1  99.0 102.2 104.9 107.2
 50.0  63.3  73.1  80.7  86.7  91.6  95.6  99.0 101.9 104.3
 50.0  61.6  70.5  77.7  83.5  88.4  92.5  96.0  99.0 101.6
 50.0  60.1  68.4  75.1  80.8  85.6  89.7  93.3  96.3  99.1
 50.0  58.9  66.5  72.9  78.4  83.1  87.2  90.7  93.9  96.7
 50.0  57.9  64.9  70.9  76.2  80.8  84.8  88.4  91.6  94.4
range=31.0, approach=39.0
```

It is important in developing such numerical solutions to differential equations to determine the appropriate grid size. The program has been written so that the grid is dynamically allocated. The number of horizontal and vertical cells was stepped from 5 to 25 by 5s and the exiting water temperature calculated to produce the following table:

Nx/Ny	5	10	15	20	25
5	89.64	89.22	89.12	89.07	89.04
10	89.21	88.97	88.92	88.89	88.88
15	89.11	88.92	88.87	88.85	88.84
20	89.06	88.89	88.85	88.84	88.83
25	89.04	88.88	88.84	88.83	88.82

The solution converges quickly with only a modest number of cells. The Runge-Kutta method for 2D may be found in Appendix C. The complete code is included in the on-line archive. This is the computational method used in the TEFERI code[25], which may be obtained through the Electric Power Research Institute (EPRI).

---

25 Bourillot, C., "TEFERI, Numerical Model for Calculating the Performance of an Evaporative Cooling Tower," EPRI CS-3212-SR, August, 1983. Originally published by Electicite de France and translated by J. A. Bartz.

Chapter 7. Natural-Draft Cooling Towers

Predicting and presenting the performance of natural-draft cooling towers is considerably more complicated than mechanical-draft because there is an additional variable: ambient relative humidity. As a result of the Merkel approximation, the performance of mechanical-draft cooling towers can be accurately represented as depending on flow, range, and ambient wet-bulb. This is not true for natural-draft towers, as the heat and mass transfer impact the density, which drives the flow of air through the tower. The impact of ambient relative humidity may even be greater than that of range or flow in some circumstances.

The classical representation of natural-draft cooling tower performance is illustrated in the next two figures. The first shows the cold-water temperature vs. wet-bulb temperature for several values of ambient relative humidity. The second is the *range correction*, that is, the adjustment in cold-water temperature for range, also vs. ambient wet-bulb temperature.

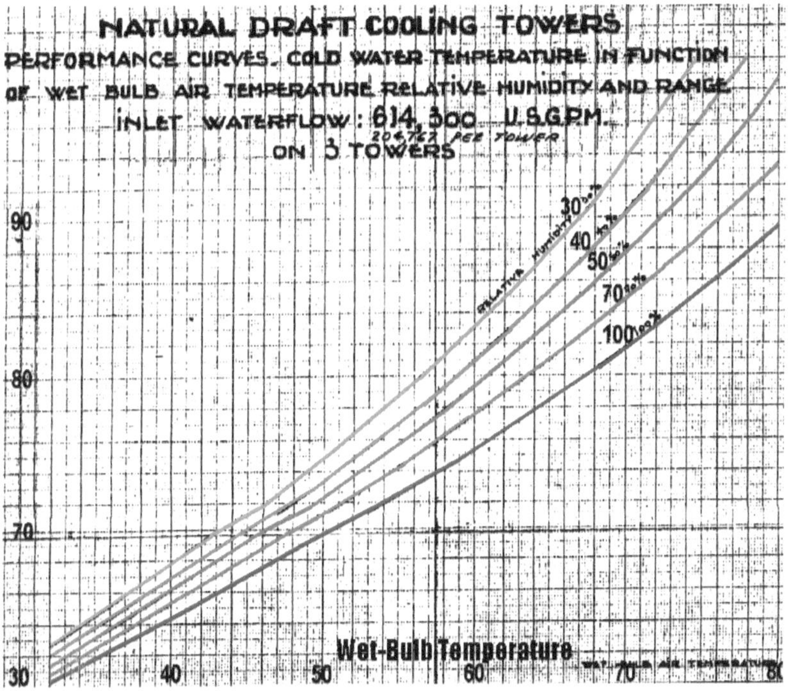

These curves are quite old (from the late 1960s), and are included here because such information is generally proprietary in nature. These curves, however, are for a U.S. government project (TVA's Paradise Steam Plant) and can be obtained through the Freedom of Information Act (FOIA).

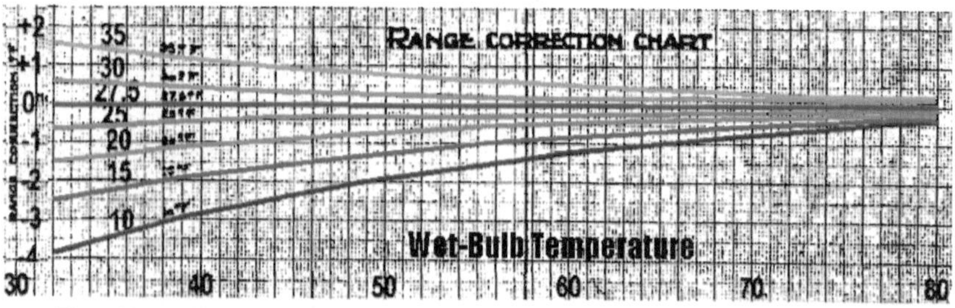

For this tower there are five sets of curves at five different water flows. These curves have been digitized and the range correction applied to each curve, resulting in over 15,000 points for cold-water temperature as a function of flow, range, ambient relative humidity, and ambient wet-bulb temperature. All of this data is included as an Excel® spreadsheet in the on-line archive. This relationship can be accurately approximated (R^2=0.9969) by a second-order multi-variable regression, as shown in this next figure:

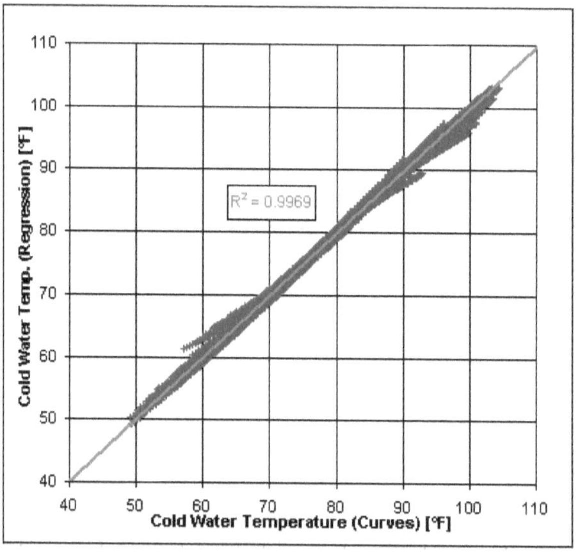

While such a regression is fully adequate for approximating the performance and correcting test data, it doesn't provide a meaningful visual presentation. A more creative presentation results from plotting curves on three sides of a cube, as illustrated in the next figure. Such a plot is usually drawn on a flat sheet of paper and then folded as indicated.

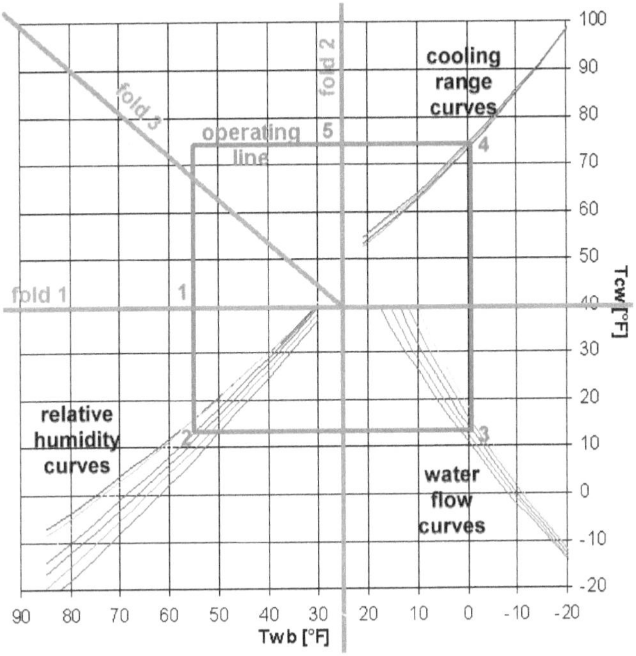

By folding along the lines as indicated above:

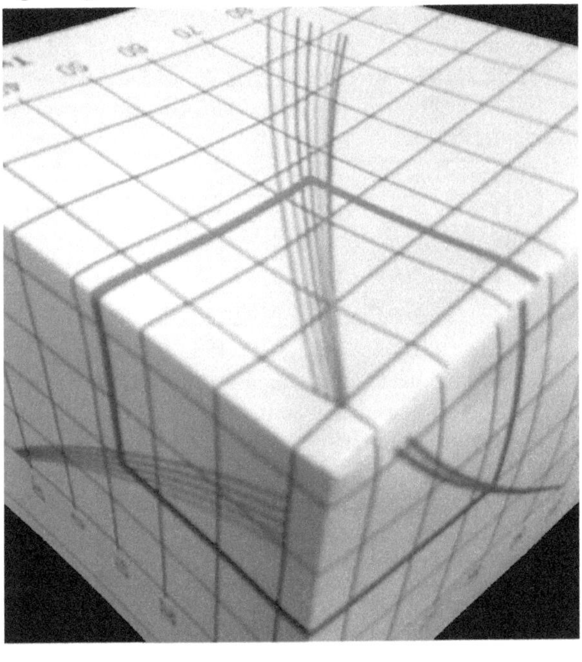

These curves representing the cooling tower performance are obtained by introducing two additional coordinates. The first will be given the symbol, ***X***, and is the distance between points 1 and 2 on the previous graph. The second will be given the symbol, ***Y***, and is the distance between points 2 and 3. These coordinates do not correspond to thermodynamic variables; they merely facilitate creation of the graph.

A typical operating point is found beginning at point 1 (the ambient wet-bulb) and drawing a vertical line down to point 2, where this intersects the ambient relative humidity curve. From point 2, a line is drawn horizontally to point 3, where it intersects the flow curve. From point 3, a line is drawn vertically to point 4, where it intersects the range curve. The cold-water temperature is then read from the scale at the right.

Each set of curves can be expressed in terms of a simple second order expansion. There are three sets of curves (relative humidity, flow, and range) and seven variables (wet-bulb, relative humidity, ***X***, flow, ***Y***, range, and cold-water temperature). Their inter-relation is given by the following equations:

$$X = a_1 + a_2 RH + a_3 Twb + a_4 RH^2 + a_5 RH \cdot Twb + a_6 Twb^2 \qquad (7.1)$$

$$Y = b_1 + b_2 flow + b_3 X + b_4 flow^2 + b_5 flow \cdot X + b_6 X^2 \qquad (7.2)$$

$$Tcw = c_1 + c_2 range + c_3 Y + c_4 range^2 + c_5 range \cdot Y + c_6 Y^2 \qquad (7.3)$$

There are eighteen constants in these three equations, but these are not all free. In order for the temperature scale of the three sets of curves to be the same, the coefficient $\boldsymbol{b_3}$ must equal one. The coefficients $\boldsymbol{a_1}$ and $\boldsymbol{c_1}$ control the horizontal and vertical gaps, or the distance from each set of curves to the corner. A regression performed on these performance curves produces the following coefficients:

a1	70	b1	-1.0061	c1	70
a2	13.77244	b2	-11.3479	c2	3.433356
a3	-1.38779	b3	1	c3	-1.26823
a4	-14.8557	b4	1.756232	c4	-0.2345
a5	0.316842	b5	-0.24559	c5	0.131833
a6	0.002149	b6	-0.00385	c6	0.006758

This provides a convenient representation of an existing tower design, but doesn't explain how the tower is designed in the first place. This particular tower was designed by Marcel LeFevre using Merkel's method. This is evident from a plot of KaV/L vs. L/G. All 15,000+ performance points lie along a downward-sloping straight line on a log-log scale, as shown in this next figure:

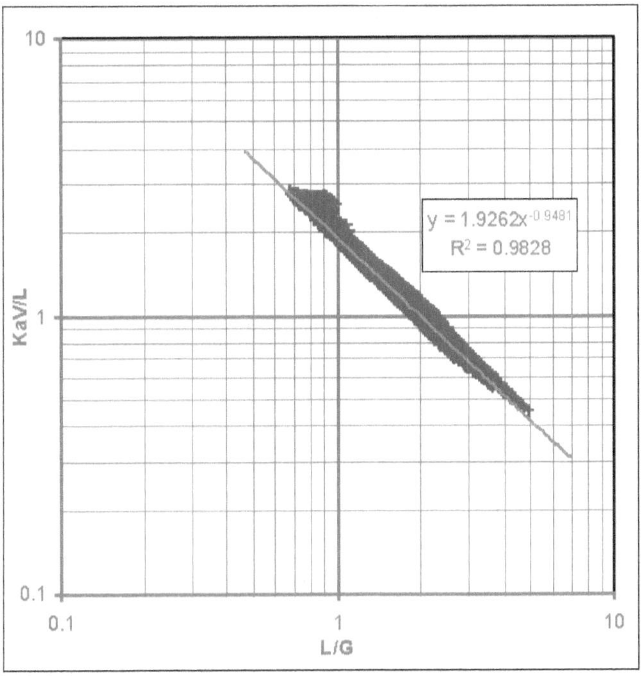

The slope and intercept are typical for the type of fill (packing) used (cement asbestos fiber board), which has since been replaced. The slope (-0.9481) is determined from small-scale test data and the intercept (1.9262) is calculated from test data and the depth of the fill (or packing), which in this design was 6 to 12 feet with an average of 11.1 feet.

This graph is still only part of the design, as the ratio of water to air mass flow rates, L/G, must be known in order to make this calculation. The exiting air from an evaporative cooling tower is saturated and at a temperature between the entering and exiting water temperatures. As a general rule, one-half may be assumed, although test data have shown a range of values between 35% and 65%. L/G can be calculated from an energy balance, by recognizing that:

$$G\left(h_{a,out} - h_{a,in}\right) = L \cdot C_{PW} \cdot \Delta T_W \qquad (7.4)$$

Equation 7.4 can be rearranged to form:

$$\frac{L}{G} = \frac{\left(h_{a,out} - h_{a,in}\right)}{C_{PW} \cdot \Delta T_W} \qquad (7.5)$$

The flow of air through a natural-draft cooling tower is driven by the density difference, which acts on the air from the top of the drift eliminators (directly above the fill and spray nozzles) to the exit of the stack. This zone is shown in the following two figures. The first figure was taken without hot water on the tower and the second was taken with.

The assumed saturated condition is clear from this second figure, which is a dense fog.

While there is some dispute in the literature concerning the appropriate boundary condition at the top of the fill and at the exit of the tower, the author has personally measured the conditions at both locations and can assure the reader that the pressure is uniform radially at both elevations. More specifically, the pressure is constant and below atmospheric above the drift eliminators. It is also constant and equal to ambient at the exit of the tower, which is typically 400 to 500 feet above the ground.

There is no sub-atmospheric pressure zone sitting on top of a natural-draft cooling tower sucking air out of it arising from the thermal plume. The pressure at the top is most assuredly atmospheric. If it weren't, there would be an inward radial flow, consistent with the Navier-Stokes equation.[26]

Recognizing that the flow of air through a natural-draft cooling tower is driven by the density difference, we can propose the following relationship:

$$\left(\frac{L}{G}\right)\Delta T_W \propto H\left(\frac{\Delta\rho}{\rho}\right) \tag{7.6}$$

In Equation 7.6, ***H*** is the height over which the density difference acts (i.e., the total height minus the inlet) and ***Δρ/ρ*** is the relative difference in density. The validity of this proposed relationship is illustrated in the next figure:

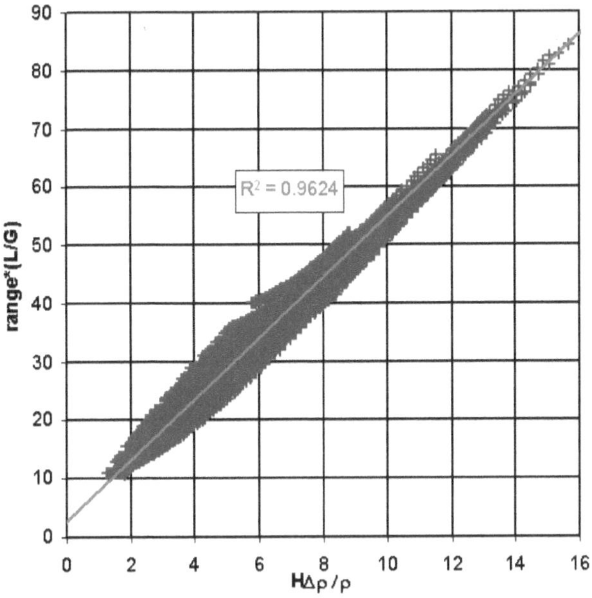

[26] Note that Bernoulli's equation only applies along a streamline, not across streamlines.

Given this linear relationship, plus the line on the previous graph of KaV/L vs. L/G, we see that only three constants (1.9262, 0.9481, 5.6337) are required to calculate all of the performance curves for this or any other natural-draft cooling tower. The slope of the ***H*** vs. ***Δρ/ρ*** line can be determined from a single solution balancing the pressure drop through the fill and drift eliminators with the buoyancy to obtain a single operating point.

<u>Dry vs. Wet Cooling</u>

There were two natural-draft cooling towers at Schmehausen, Germany (pictured below). The smaller one on the left (made of concrete) is wet and the larger one on the right (made of steel) is a dry. Both were in operation when this picture was taken.

It cost ten times as much to construct the large dry cooling tower than the smaller wet cooling tower next to it. Besides being much smaller and cheaper, the wet cooling tower provided *far better performance*, hence the advantage of evaporative cooling. This dry cooling tower was demolished in 1991, because the materials were worth more as scrap than it took to build another wet tower.

Chapter 8. Heat Transfer from Falling Droplets

Evaporative heat transfer from falling droplets is an important industrial process. The *spray zone* in a cooling tower is only one such application. In this chapter we will consider this process from both a computational and empirical basis. This chapter is largely a condensation of a paper presented by the author and Bob Rehberg in 1986, which contains analysis and laboratory data.[27]

In a classic reference, Lowe & Christie[28] discuss a method developed by Nottage & Boelter[29] for calculating the mass transfer characteristic and pressure drop of droplets falling in a counterflow arrangement. The mass transfer characteristic, ***Ka/L"***, of the droplets is defined as:

$$\frac{Ka}{L"} = \left(\frac{C_{PL}}{C_{PG}}\right)\left(\frac{\kappa_G}{\kappa_L}\right)\left(\frac{\Psi}{s_G}\right) \tag{8.1}$$

and has units of 1/length. In this equation C_{PL} and C_{PG} are the liquid and vapor specific heats of water, respectively, κ_G and κ_L are the thermal conductivity of the liquid and vapor, respectively, and S_G is the velocity of air.

The *dynamical function*, Ψ, defined and tabulated by Nottage and Boelter, is a function of droplet diameter and has units of 1/time. The relative velocity of the falling droplet, s_A, is the difference between the free falling or terminal velocity, s_F, and the air velocity, s_G. The terminal velocity is a function of droplet diameter and is tabulated in the reference by Nottage and Boelter. The air velocity through the tower is defined as:

$$S_G = \frac{G"V_A}{3600} \tag{8.2}$$

which has the units of length/time. The pressure drop per length l is given by:

$$dP = \left(\frac{\rho g L"}{3600\, s_A}\right) dl \tag{8.3}$$

and has the units of force/length². The pressure drop, as expressed in the number of velocity heads, ***dN***, lost per foot is:

---

27 Benton, D. J. and R. L. Rehberg, "Mass Transfer and Pressure Drop in Sprays Falling in a Freestream at Various Angles," International Association for Hydraulic Research, Fifth Cooling Tower Workshop, Palo Alto, California, September 29 - October 3, 1986.

28 Lowe, H. J., and D. G. Christie, "Heat Transfer and Pressure Drop Data on Coolïng Tower Packings and Model Studies of the Resistance of Natural Draught Towers to Airflow," International Division of Heat Transfer, Part V, pp. 333-950, American Society of Mechanical Engineers, New York, 1961.

29 Nottage, H. B., and L. M. K. Boelter, "Dynamic and Thermal Behavior of Water Drops in Evaporative Cooling Processes," Transactïons of the American Society of Heating and Ventilating Engineers, Vol. 46, pp. 41-79, 1940.

$$\frac{dN}{dl} = \left(\frac{2}{s_G^2}\right)\frac{dP}{dl} \qquad (8.4)$$

and has the units of heads/length. Substitution of Equation 8.3 into 8.4 yields:

$$\left(\frac{1}{L''}\right)\frac{dN}{dl} = \frac{2g}{3600\,\rho\, s_A s_G^2} \qquad (8.5)$$

and has the units of length×time/mass. This development by Lowe & Christie has been used for years to model the heat transfer in the rain zone of cooling towers. A more general approach based on first principles and dimensionless correlations will be developed here that could also be used for other substances and orientations.

Theoretical Development

The mass transfer is determined from the Sherwood number, ***Sh***:

$$Sh = \frac{Kad}{\rho_A\, D_{AW}\, A_V} \qquad (8.6)$$

where $\boldsymbol{\kappa_A}$ is the thermal conductivity, $\boldsymbol{d}$ is the droplet diameter, $\boldsymbol{\rho_A}$ is the density of the air, $\boldsymbol{D_{AW}}$ is the air/water diffusion coefficient, and $\boldsymbol{A_V}$ is the surface area per unit volume. For flow over a sphere, the Sherwood number is given by:[30]

$$Sh = 2 + \left(0.4\,\mathrm{Re}^{\frac{1}{2}} + 0.6\,\mathrm{Re}^{\frac{2}{3}}\right) Sc^{0.4} \qquad (8.7)$$

where ***Re*** is the Reynolds number and ***Sc*** is the Schmidt number.

$$\mathrm{Re} = \frac{V_R \rho_A d}{\mu_A} \qquad (8.8)$$

where $\boldsymbol{V_R}$ is the velocity of the air relative to the droplet and $\boldsymbol{\mu_A}$ is the dynamic viscosity of air.

$$Sc = \frac{\mu_A}{\rho_A D_{AW}} \qquad (8.9)$$

The surface area per unit volume, $\boldsymbol{A_V}$, is given by:

$$A_V = \frac{A_S \phi_D}{V_I} \qquad (8.10)$$

where $\boldsymbol{A_S}$ is the surface area per droplet in length²/drop, $\boldsymbol{\varphi_D}$ is the flux of droplets in drops/length²/time, and $\boldsymbol{V_I}$ is the instantaneous droplet velocity. The relative velocity, $\boldsymbol{V_R}$, in Equation 8.8 is given by:

[30] Krieth, F. and W. Z. Black, Basic Heat Transfer, Harper and Rowe, 1980.

$$V_R = \sqrt{(u_A - u_D)^2 + (v_A - v_D)^2} \tag{8.11}$$

The heat and mass transfer for a single drop is computed from the Lagrangian Equations of Motion.[31] The horizontal and vertical position of a particle is given by:

$$x = \int_0^t u dt \tag{8.12}$$

$$y = \int_0^t v dt \tag{8.13}$$

The horizontal and vertical velocities are given by:

$$u = \int_0^t \frac{F_X}{m_D} dt \tag{8.14}$$

$$v = \int_0^t \frac{F_Y}{m_D} dt \tag{8.15}$$

The horizontal and vertical impulse are given by:[32]

$$J_X = \int_0^t F_X dt \tag{8.16}$$

$$J_Y = \int_0^t F_Y dt \tag{8.17}$$

and have the units of force×time. The horizontal force, $\boldsymbol{F_X}$, is entirely due to *drag* and is given by:

$$F_X = \left(\frac{\pi d^2}{4}\right) C_D \frac{\rho_A |u_A - u_D| (u_A - u_D)}{2} \tag{8.18}$$

The vertical force, $\boldsymbol{F_Y}$, is due to drag and gravity and is given by:

$$F_Y = \left(\frac{\pi d^2}{4}\right) C_D \frac{\rho_A |v_A - v_D| (v_A - v_D)}{2} - (\rho_D - \rho_A) g \frac{\pi d^3}{6} \tag{8.19}$$

---

[31] The Lagrangian approach follows the perspective of individual particles along their trajectories. This is in contrast to the continuum approach of Euler, most often used for fluids. This approach was introduced by the Italian-French mathematician and astronomer Joseph-Louis Lagrange in 1788.

[32] In classical mechanics, impulse is the integral of force over the time interval for which it acts. Impulse applied to an object produces a change in its linear momentum. Linear momentum is equal to mass times velocity. Force, impulse, and linear momentum are all vectors.

The difference in densities in the last term accounts for the buoyancy. The pressure drop across the zone is found by integrating Equations 8.12 through 8.17 along with Equation 8.4 to obtain:

$$NV' = \frac{\phi_D \sqrt{I_X^2 + I_Y^2}}{H\left(\frac{\rho_D V_D^2}{2}\right)} \tag{8.20}$$

where ***H*** is the height of the zone. The force balance on a single drop is illustrated in the following figure:

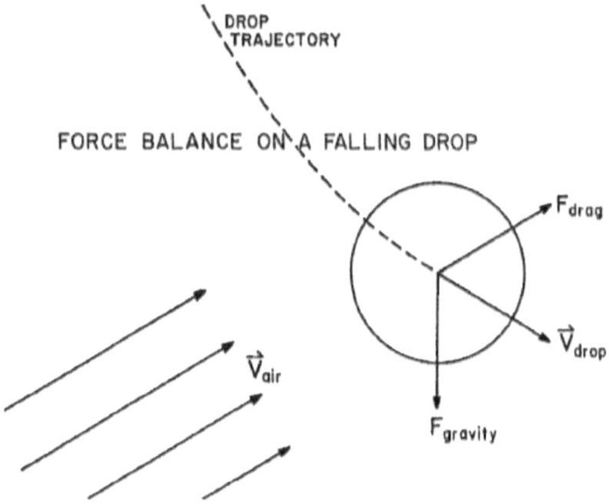

The program used to solve these equations and produce the graphs in the next section is listed in Appendix D.

Comparison to Experimental Results

There have been several experimental studies done on the mass transfer and pressure drop from single droplets falling through air. One study conducted by Missimer & Brackett at the TVA Engineering Laboratory in Norris, Tennessee, provides experimental data specifically on the transport phenomena occurring within the rain zone of a counter flow cooling tower. The test facility was designed to simulate the rain zone in the air flows toward the center of the tower in a crossflow configuration.[33]

The test section is 15 feet long, 6 feet high and 4 feet wide with a maximum water flux of 2085 lbm/hr/ft² and a maximum air flow rate of 15,000 ft³/min. The parameters measured during the experiments were the air and water flow rates, the hot water temperature entering the test section, the cold water temperature leaving the test section and the inlet wet bulb temperature. Given

[33] Missimer, J. R., and C. A. Brackett, "Results of Model Tests of Heat Transfer in the Rain Zone of a Counterflow Natural-Draft Cooling Tower," TVA Report WR28-1-85-115, May 1985.

this data, the mass transfer characteristic, Ka/L", can be computed. Also, the pressure drop across the test section was measured, thereby giving the number of velocity heads lost per foot of air travel, ***NV'***.

The results of the initial 12 tests conducted at the TVA Rain Zone Facility are given in the following figure. In addition to the experimental data, the results of the present numerical analysis are also shown.

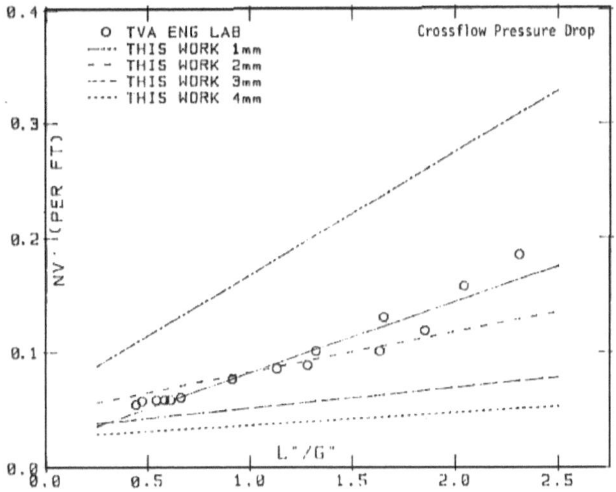

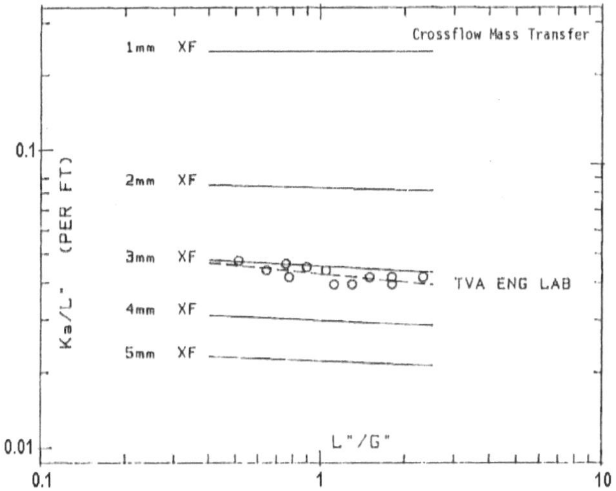

The theoretical curves show the dramatic effect of droplet size on the mass transfer. From this figure, it can be seen that a difference in droplet diameter of a few millimeters can produce an order of magnitude difference in the mass transfer characteristic. Accurate droplet sizes were not experimentally determined during the TVA rain zone testing. However, the theoretical curve for 3-mm droplets compare well with the experimental data, which is roughly the correct droplet diameter.

No experimental data for counterflow mass transfer were available. Two sets of curves are shown in the figure: the solid curves were computed based on the analysis discussed by Lowe and Christie, while the dashed curves were computed using the equations detailed above. As with the case of crossflow mass transfer, the dramatic effect of droplet size can be seen.

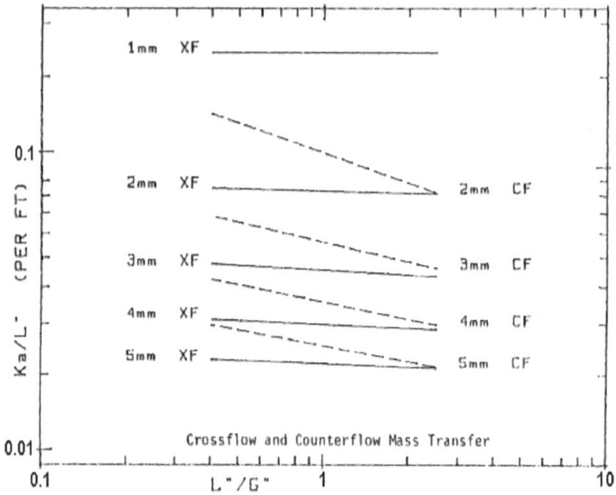

The figure also shows the decrease in mass transfer coefficient w1th increasing water loading (increasing ***L"/G"***). An interesting comparison of crossflow and counterflow ra1n zone mass transfer is also shown in this figure. As expected, for a given droplet diameter the mass transfer coefficient is higher for the counterflow orientation.

Since the rain zone in a natural-draft counterflow cooling tower is a combination of both counterflow and crossflow, for a given droplet size, the mass transfer characteristic for the rain zone would presumably lie between the counterflow and crossflow curves. Therefore, the curves shown in this figure provide a theoretical upper and lower bound on the mass transfer actually occurring in the rain zone. Typical computed droplet trajectories are shown in this next figure:

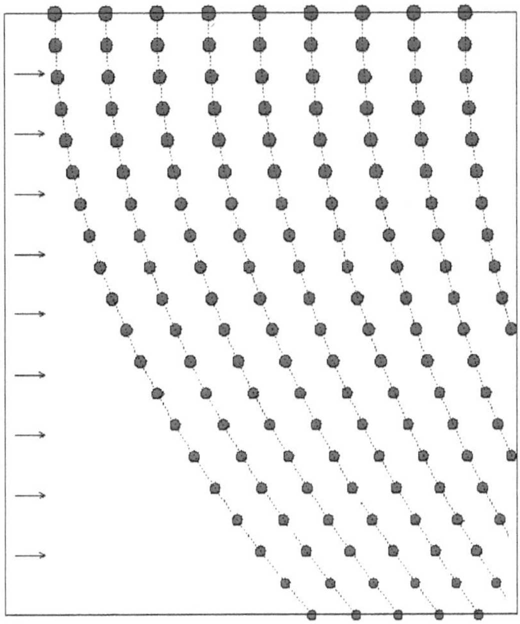

Chapter 9. Spray Cooling Systems

Large-scale open spray cooling systems were very popular in the early 1970s for their simplicity, reliability, and low cost. Such systems were even proposed and accepted as the ultimate heat sink for several nuclear plants. An Internet search will produce several U.S. Nuclear Regulatory Commission (NRC) documents containing descriptions and analyses for these systems from this era, but very few more recent publications.

By the late 1970s and early 1980s these systems came under attack by environmental activists, concerned that they were releasing harmful substances and/or radiation into the atmosphere, thereby endangering the public. While there was no basis for these concerns, spray systems are more visible, giving the false impression that something more significant is happening.

The spray cooling pond at the Rostov Nuclear Power Plant in Volgodonsk, Russia is shown in this next figure:

It can be challenging to quantify the cooling achieved by such systems. Fortunately, the supporting analysis has been published.[34,35] Frohwerk's patent contains an excellent description of spray systems.[36] The airflow induced by the spray process carries away the absorbed energy in the form of sensible and

---

[34] Berger, M. H. and Taylor, R. E., "An Atmospheric Spray Cooling Model," Proceedings of the 2nd AIAA/ASME Thermophysics and Heat Transfer Conference, Palo Alto, California, May 24-26, pp. 59-64, 1978.

[35] Chaturvedi, S. K. and R. W. Porter, "Thermal Performance of Spray-Canal Cooling Systems," Journal of Engineering for Power Vol. 102, No. 4, pp. 776-781, 1980. (available free on-line, search for 19770077431.pdf)

[36] United States Patent No. 3,622,074 issued to Paul A. Frohwerk of the Ceramic Cooling Tower Company on November 23, 1971. (available free on-line)

latent heat. The airflow and heat transfer are intimately linked. The conservation of energy can be expressed as follows:

$$G\Delta h = LC_{P}\Delta T_{W} \tag{9.1}$$

The Lagrangian droplet tracking model described in the last chapter can be run for a range of droplet diameter, ***d***, and falling height, ***h***, to produce the following graph of 1-effectiveness, or approach/(range+approach):

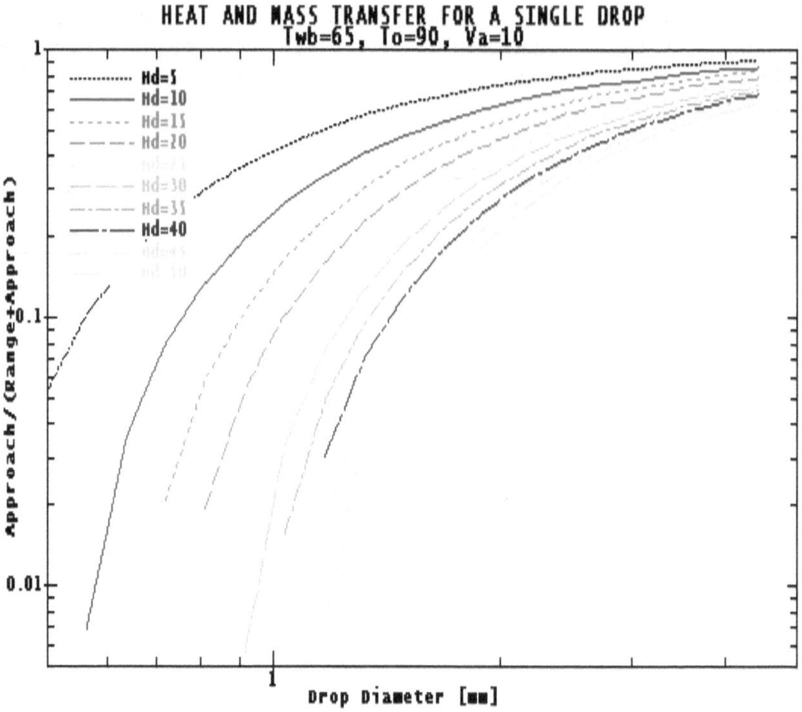

Based on regression, these curves can be approximated by the following formula for d≥0.5mm:

$$\varepsilon = 1 - \left\{ e^{[a + b\ln(h) + c\ln(d)]} - 1 \right\}^{3} \tag{9.2}$$

where a=-2.45591150714, b=0.631744225842, and c=-1.36867783072. As indicated at the top of the figure, this is for a single droplet. There are several manufacturers of high-quality industrial spray nozzles, including: Bete®, Lechler®, Phirex®, and Spray Systems®. The following is a Phirex® FogEx® nozzle used in fire suppression systems:

Spray nozzles produce a distribution of droplet sizes. The following distribution of droplet sizes is for a Bete® full-cone nozzle with 7 psig operating pressure.

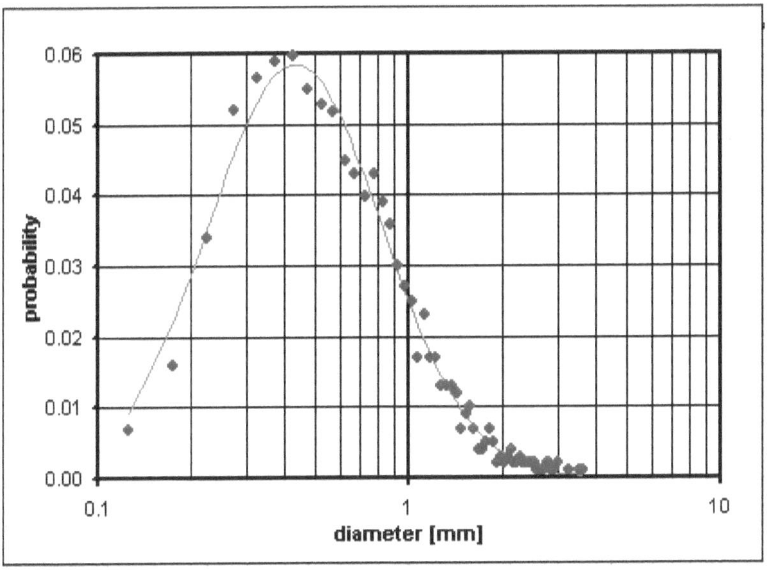

The blue dots are measured data and the red curve is a lognormal distribution having a mean of 0.43 mm and a standard deviation of 1.9 mm. The droplet size distribution can be applied to the preceding effectiveness curves[37] to arrive at a single curve of effectiveness vs. height for the droplets created by this nozzle at this operating pressure. This curve only applies to an average single droplet.

---

[37] The performance of the distribution of drop sizes is equal to the sum of the probability times the performance for each drop size, $\underline{\varepsilon}=\sum Þ\varepsilon/\sum Þ$. The sum of the probabilities is equal to one.

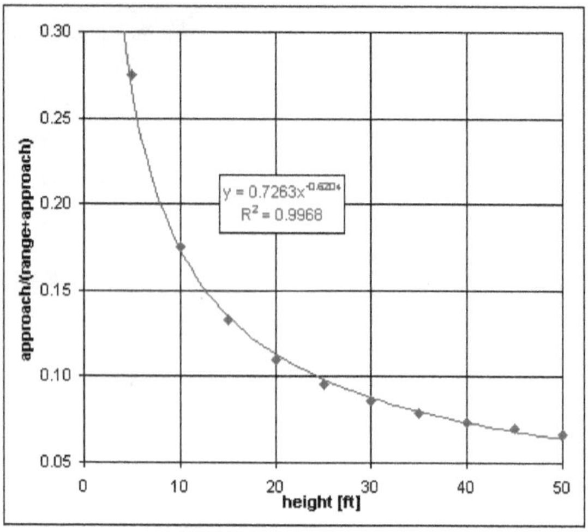

As water vapor (molecular weight ≈18) is lighter than air (molecular weight ≈29), evaporation from the spray will induce an upward draft of air, even in the absence of a crosswind. In the case of an open spray, as in the typical power plant application depicted here, the air flows radially inward and then upward. The streamlines formed by this flow may be approximated by the potential flow produced by two point sinks at $\pm h$, where h is some characteristic height. This forms a plane of symmetry at the ground, where there is no vertical velocity. This flow field is illustrated in the following figure:

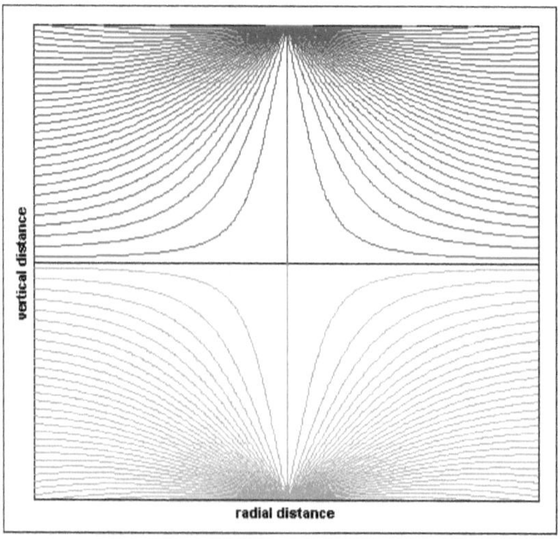

Floating spray modules have been deployed for decades in ponds and channels for supplemental cooling and aeration at steel mills, pulp mills, and wastewater plants. Utilization of these systems at power plants has been less common, at least in the U.S. Such systems can be quite cost-effective and there are still a few manufacturers, including ARWADH, whose 75 hp ThermoFlo® system is shown below (from their on-line brochure):

It is fortuitous that the same relationship for heat transfer and draft that governs the flow of air through a natural-draft cooling tower holds for a falling spray. This is particularly useful, as the problem of modeling this type of flow is much more complicated. This is primarily due to the fact that the flow in a cooling tower is confined by the structure; whereas, the flow induced by a spray is not. The fill also serves to guide the flow through the tower structure.

The following figure shows temperature data collected for the spray cooling pond at a paper mill:

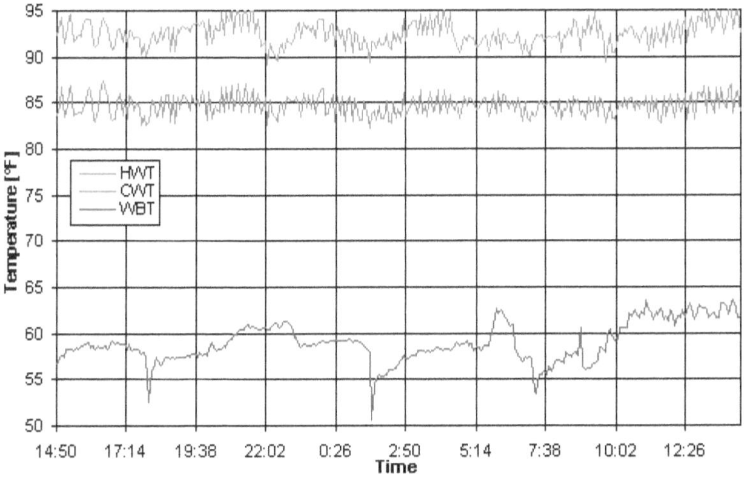

The water-to-air mass flow ratio, ***L/G***, can be calculated from Equation 9.1 and the mass transfer characteristic, ***KaV/L***, can be calculated from Equation 3.1. This next figure shows the results of these calculations:

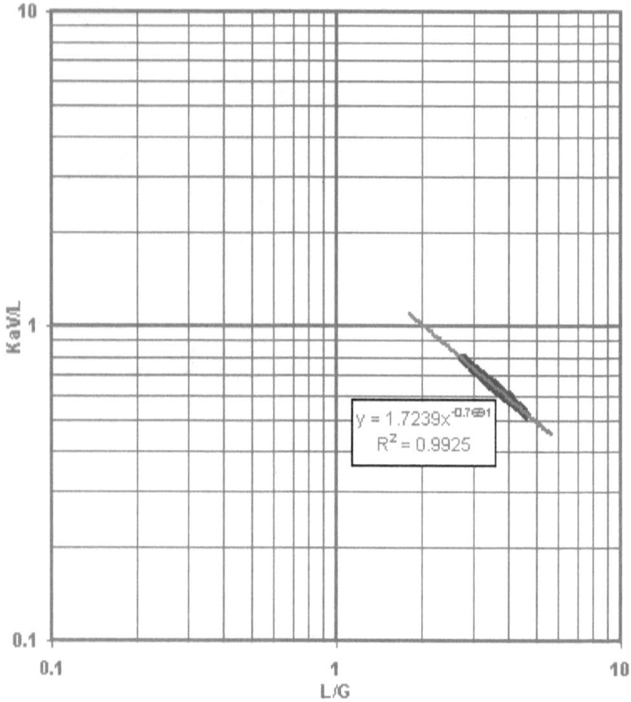

Data for this spray cooling system produces a remarkably tight cluster (R^2=0.9925) about the line ***KaV/L=1.7239/(L/G)***$^{0.7691}$, clearly illustrating the facility of Merkel's theory and this approach.

Chapter 10. Cooling Ponds

There was a great deal of interest in large cooling ponds during the early days of nuclear power plant design and construction in the United States. The U.S. Nuclear Regulatory Commission (USNRC) and the U.S. Environmental Protection Agency (USEPA) funded a series of studies extolling the efficacy of such ponds for waste heat removal.[38,39,40,41] Some of these articles may be found on-line. Several of these are included in the archive that accompanies this text.

Most of these studies refer back to the work of Langhaar.[42] While this is a very old reference, the properties of air and water have not changed. Langhaar's analysis was quite insightful and his calculations continue to be relevant and useful. This method has been updated and refined for presentation here. All of the calculations have been implemented in an Excel® spreadsheet that includes actual meteorological data and the performance of a certain nuclear plant that utilizes a lake for cooling.

All cooling pond calculations begin with the definition of an equilibrium temperature. This is the temperature a pond will eventually approach under steady conditions. The following figure is based on Langhaar's nomograph:

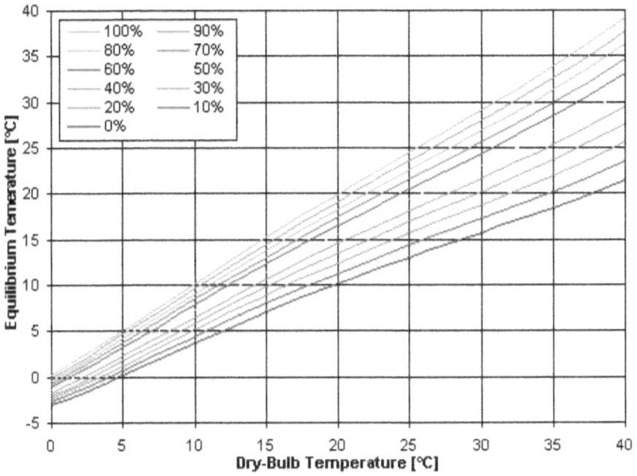

---

[38] Hogan, W. T., Liepins, A. A., and Reed, F. E., "An Engineering - Economic Study of Cooling Pond Performance," EPA Project 16130DFX05/70 Contract No. 14-12-521, May, 1970.

[39] Hadlock, R. K. and Abbey, O. B., "Thermal Performance Measurements on Ultimate Heat Sinks - Cooling Ponds," NUREG/CR-0008, February 1978, also published as a Batelle Pacific Northwest Laboratories Report PNL-2463.

[40] Codell, R. and Nuttle, W. K., "Analysis of Ultimate Heat Sink Cooling Ponds," NUREG-0693, November, 1980.

[41] Berger & Taylor include some helpful discussion as well (see Reference 34).

[42] Langhaar, J. W., "Cooling Pond Many Answer Your Water Cooling Problem," Chemical Engineering, August, 1953, pp. 194-199.

It is important to note that the equilibrium pond temperature is not simply the web-bulb; rather, it is based on empirical data. The following figure shows the difference between the equilibrium and web-bulb temperatures:

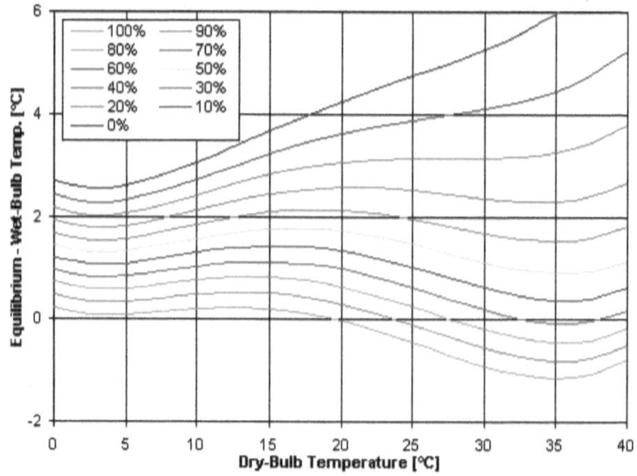

Wind and solar are strong influences on pond temperature, so that any model must contain corrections for these factors. The following figure shows the impact of wind speed on the equilibrium temperature with a Global Horizontal Irradiance (GHI) of 750 W/m². There are similar plots for 1000, 500, and 250 in the spreadsheet named Langhaar.xls. The functions to perform these calculations are also included as macros.

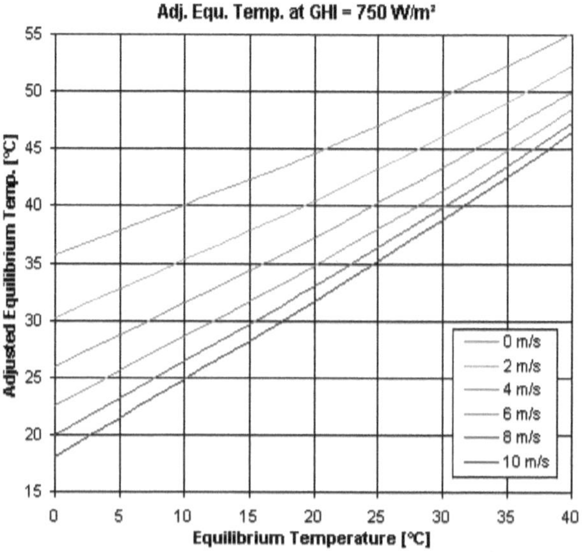

Langhaar's method includes two more parameters related to scale, given the symbols ***P*** and ***Q***. The product ***PQ=A/F*** is the area divided by flow and is dimensional so that unit conversions are necessary for implementation. This has been implemented in the example spreadsheet, cooling_pond_simulation.xls, which models the performance of the nuclear plant and contains all of the necessary scaling factors.

The best way to illustrate Langhaar's model of pond performance is through an actual application where a large-scale pond is used to cool a power plant, in this case nuclear. The thermal performance of the plant is needed in order to capture the response of the heat rejection to the ambient conditions.

Performance functions for a particular nuclear plant is provided in Appendix E. This performance is typical of all nominal 1000 MWe Westinghouse pressured water reactors. Functions for the pond performance are provided in Appendix F. Functions that combine these two in order to model the response of the plant and pond are provided in Appendix G. All of this code is included in the on-line archive.

Pond area is determined from bathymetric surveys. Regression is used to fit area vs. elevation. Volume is determined by integrating the area with respect to elevation from the bottom up. The pond geometry is shown in this next figure:

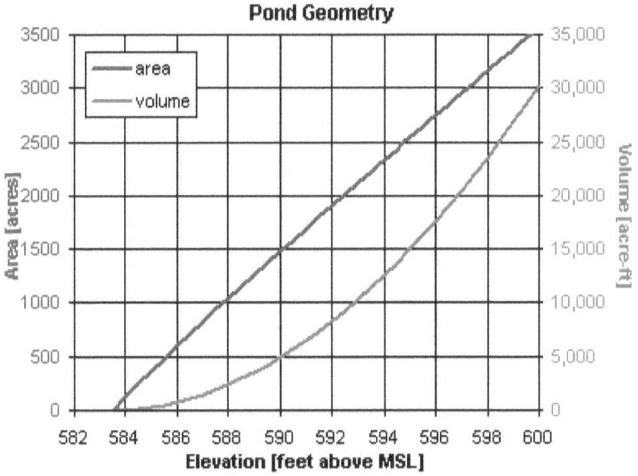

Hourly meteorological data was obtained from the National Climate Data Center. The dry-bulb and dew point temperatures are shown in the following figure:

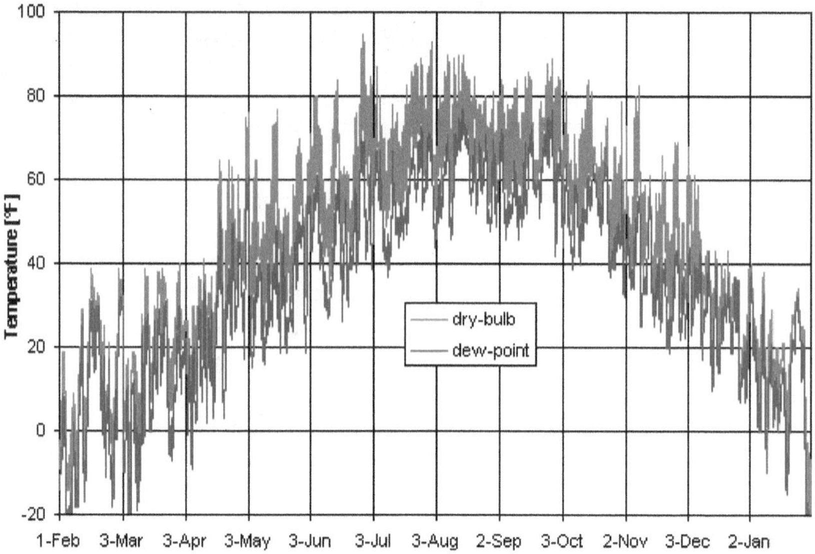

The Direct Normal Irradiance (DNI) and Global Horizontal Irradiance (GHI) is shown in this next figure:

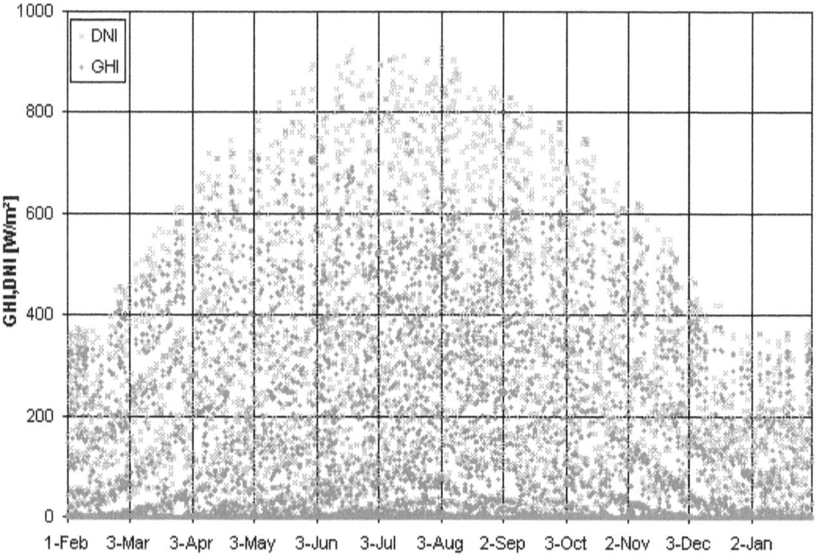

The calculated entering and exiting pond temperature is shown in this next figure:

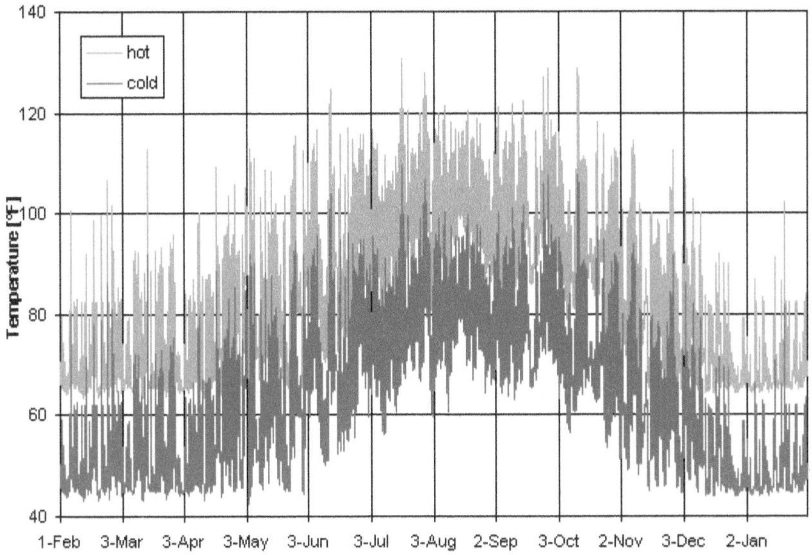

The generator output and zone three (highest) condenser pressure is shown in this next figure:

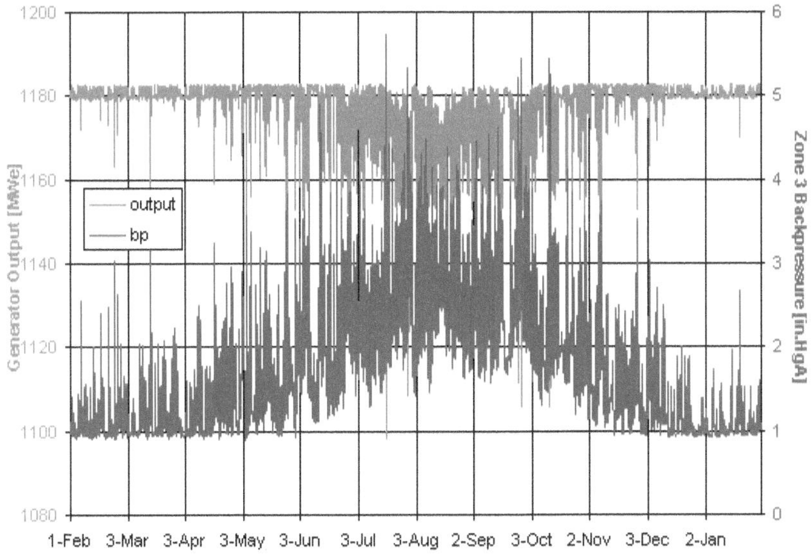

The simulation inputs and outputs are included in the archive in files nuke_input.csv and nuke_output.csv, respectively.

Chapter 11. The Lewis Number

In the derivation of Merkel's Equation, it was assumed that the coefficient of sensible heat transfer was proportional to that of mass transfer coefficient. This assumption was represented by Equation 3.8. The Lewis number is often expressed as:

$$L_E = \frac{Sc}{\Pr} \tag{11.1}$$

where ***Sc*** is the Schmidt number and ***Pr*** is the Prandtl number. The Schmidt number is equal to:

$$Sc = \frac{\mu}{\rho D} \tag{11.2}$$

where $\boldsymbol{\mu}$ is the dynamic viscosity, $\boldsymbol{\rho}$ is the density, and ***D*** is the diffusion coefficient or mass diffusivity. The Prandtl number is equal to:

$$\Pr = \frac{\mu C_P}{\kappa} \tag{11.3}$$

where $\boldsymbol{C_P}$ is the constant pressure specific heat and $\boldsymbol{\kappa}$ is the thermal conductivity. Equations 11.1 through 11.3 can be combined to form:

$$L_E = \frac{\kappa}{\rho C_P D} \tag{11.4}$$

This expression of the Lewis number is a combination of molecular properties, thus, independent of any process peculiarities. This may be more accurately termed the *molecular* Lewis number, while Equation 3.8 may be more accurately termed the *turbulent* Lewis number. This ratio of properties for air is shown in the following figure:

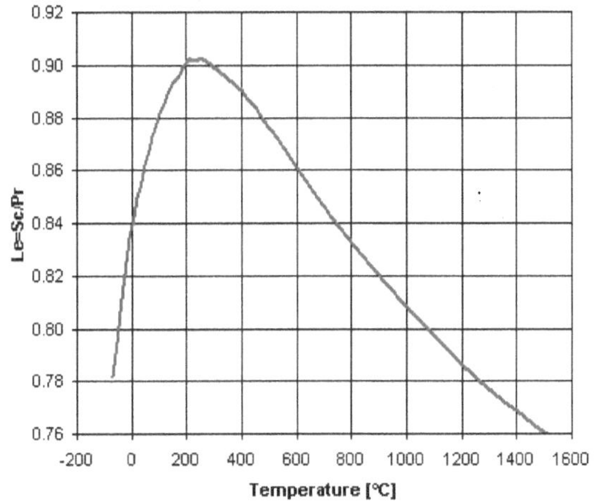

The molecular Lewis number for air never reaches unity, topping out at 0.902. At the conditions most often found in evaporative cooling, this ratio of properties is between 0.85 and 0.88. The turbulent Lewis number is of greater interest in evaporative cooling, but this can be difficult to determine and can't be directly measured, as the sensible and latent transfer processes can't be physically separated.

Little has been published on this subject, especially in the area of evaporative cooling, although there is at least one publication addressing this issue.[43] The method presented in this reference is a simple inverse heat or mass transfer problem with the two unknowns being the mass transfer coefficient and the sensible heat transfer coefficient. The ratio of these and the specific heat yields the experimental Lewis number.

It is easy enough to set up the problem and solve for the unknowns by minimizing the resulting errors. In this case the errors are the difference between the calculated and measured exiting water and air temperatures. As it turns out, if the exiting air is entirely saturated, the solution is degenerate, so the data set must be carefully selected to avoid this condition.

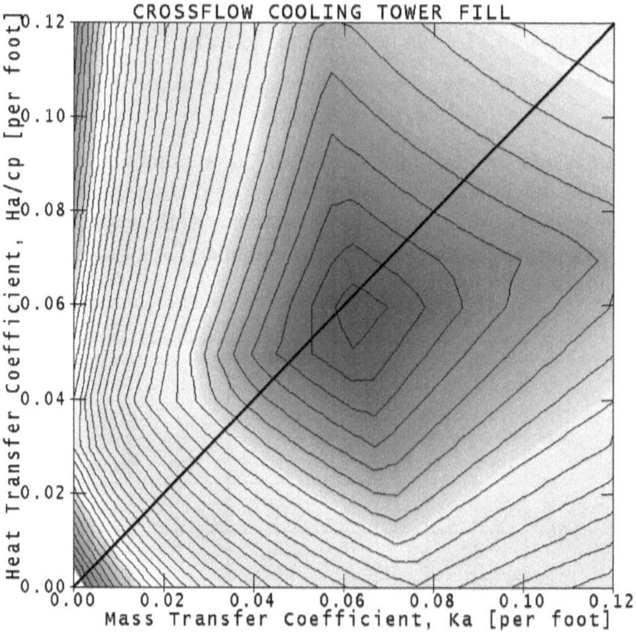

[43] Benton, D. J., "Determination of the Turbulent Lewis Number from Experimental Data for Wet Cooling Tower Fills," Cooling Tower Institute TP90-07, 1990.

Perhaps even more interesting than the actual value of the experimental Lewis number is the sensitivity of the residuals to the unknown parameters. This peculiarity was only discovered by accident during the debugging of the software. The residual is given by:

$$r = \sqrt{(\Delta T_W)^2 + (\Delta T_A)^2} \tag{11.5}$$

Contours of this residual can be generated by varying the unknown coefficients over a range of values. Ideally, this would look something like the preceding figure. Small values of the residual are indicated by blue and large values by red. The actual magnitude of the residuals is immaterial. The smallest residual is located at the center of the roughly circular black contour at X=0.065, Y=0.055. This ratio indicates an experimental Lewis number of approximately 0.85, which is in line with the molecular value.

The diagonal black line corresponds to a Lewis number of unity. The center of this contour lies just below this line. This is a cross flow case and is reminiscent of a bull's-eye. In three dimensions this residual map would have the shape of a bowl.

As it turns out, this figure is not at all typical. In fact, it took some time to find a case within the data set that produced such a residual map. The following is more typical, but still ideal. It is a counter flow case. The contour having the smallest residual (deepest blue) is centered on the diagonal line, indicating an experimental Lewis number of 1.

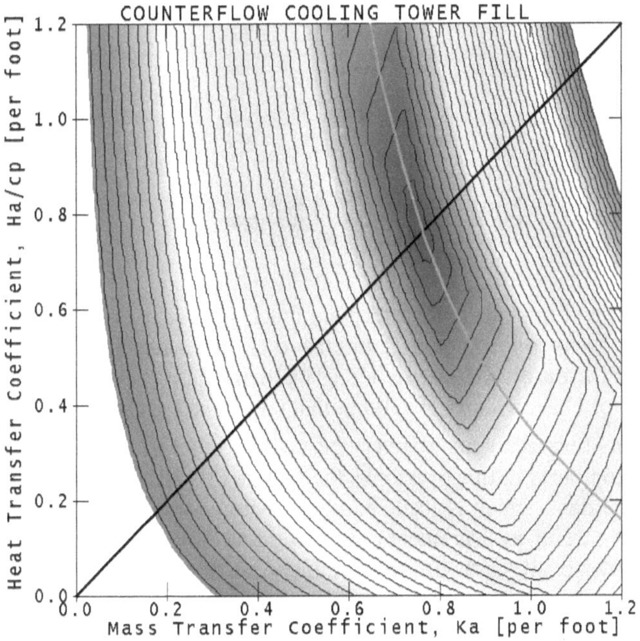

The optimum solution to this inverse combined heat and mass transfer problem corresponds to the smallest residual. This is a specific case of the more general problem of nonlinear minimization. The physical analogy to this mathematical process is one of finding the lowest elevation on a topographic surface. In this case, the longitude and latitude correspond to the two transfer coefficients, ***Ka*** and ***Ha/Cp***, and the elevation corresponds to the residual.

This figure shows a very interesting character of this problem. Every value of X and Y inside a contour has the same or lower residual, making these equally valid solutions. The best solutions (i.e., the ones having the smallest residuals) lie along the magenta curve. In three dimensions this residual map would not have the shape of a bowl, rather it would be more like a gully, as shown in this next figure:

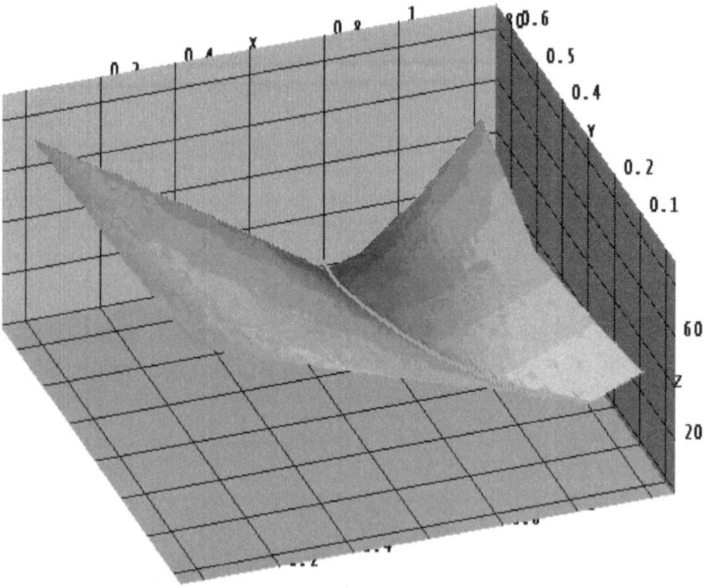

The sides of this residual surface are much steeper in one direction than the other. The locus of optimum solutions (i.e., the ones having the smallest residuals or lowest elevations) lie along the magenta arc. All solutions with the same shade of blue are equally valid in that they have the same residual.

It is tempting to assume the very best solution lies at the center of the inner most blue contour, but this overlooks the fact that these results are derived from experimental data, which is itself burdened with some level of uncertainty. No matter how careful measurements are made and how recently instruments are calibrated, these all have limited accuracy so that the data are not known exactly.

As seen in the two previous figures, the residual surface is much steeper across the magenta arc than it is a along the arc. The significance of this is that, while there may be considerable uncertainty as to the optimum distance along the arc, there is much less uncertainty as to the optimum distance perpendicular to the arc. For instance, given the accuracy of the data, we may be able to determine the distance along the arc to within ±5% and the distance perpendicular to the arc to within ±1% if the contours have an aspect ratio of 5:1.

The shape of the residual surface and contours of this particular data set (cooling tower packing of the *splash* type) vary considerably, as does the aspect ratio. Several of the cases had an aspect ratio of 50:1 or more, which is analogous to a very deep canyon. This is presents a significant problem when analyzing the data. This next figure is similar to many of the cases analyzed.

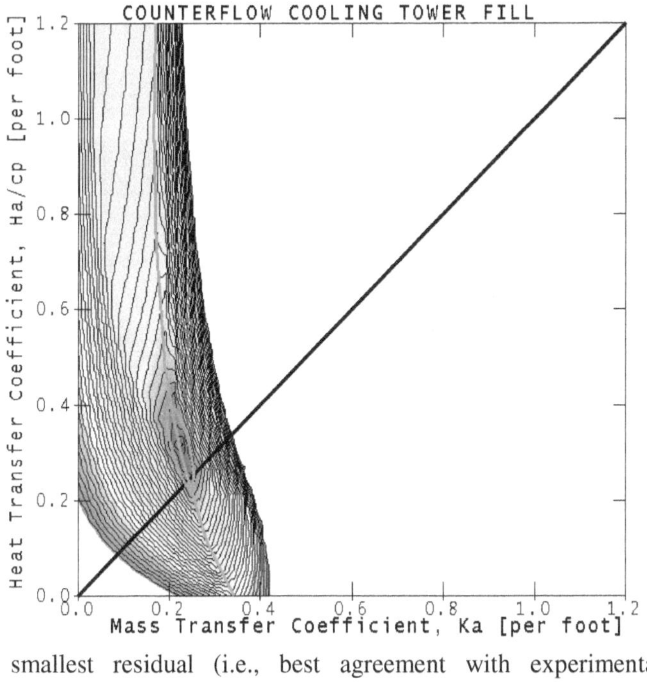

The smallest residual (i.e., best agreement with experimental data) corresponds to approximately ***Ka=0.25*** and ***Ha/C$_P$=0.30*** for an experimental Lewis number of ***L_E=1.2***. Considering the uncertainty in the test data, the solution could be anywhere inside the magenta ellipse, that is ***$0.2 \le Ka \le 0.3$*** ***$0.25 \le Ha/C_P \le 0.4$***, such that the uncertainty of ***Ha/Cp*** is greater than ***Ka***.

EPRI built a small-scale test facility for cooling tower fill at Houston Lighting and Power's Parish Station. Several counter flow and cross flow packings were tested and the results published in an EPRI report.[44] These data were used to determine the experimental Lewis number as described in TP90-07. The following figures show the cross flow and counter flow data, respectively:

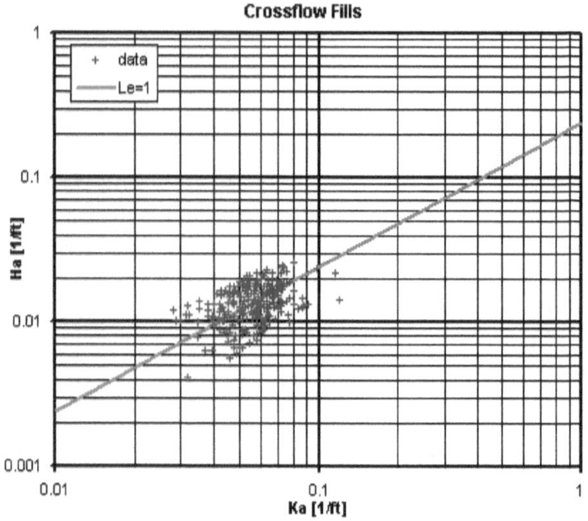

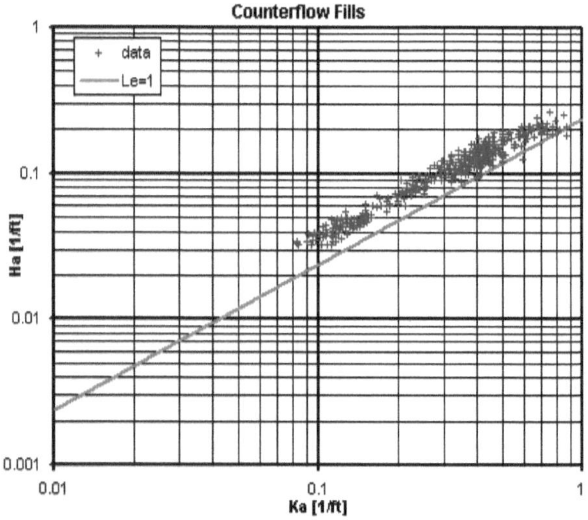

---

[44] Bell, D. M., B. M. Johnson, E. V. Werry, P. B. Miller, D. E. Wheeler, K. R. Wilbur, D. J. Benton, and J. A. Bartz, "Cooling Tower Performance Prediction and Improvement," EPRI RP2113 1988.

There is considerable scatter in the cross flow graph. The average experimental Lewis number for this data is 1.03±0.59 and for the counter flow is 1.40±0.37. It is therefore reasonable to use a value of unity for cross flow analysis, but not for counter flow. Both the cause and significance of this difference is still unknown.

Chapter 12. Air into Water

It is truly remarkable that the same approach presented in Chapter 8 for water droplets evaporating into air can be used to analyze air bubbles absorbing into water. This final chapter has been included to give perspective on the evaporative cooling process within the greater context of mass transfer. This discussion and data are based on a report written for the U. S. Army Corps of Engineers (USACE).[45]

The goal of the study presented in this report was to develop and utilize a numerical model of the nitrogen supersaturation process occurring in the plunge pool beneath dam at the Jennings Randolph Lake Project. A literature review was conducted and the model developed by the USACE Waterways Experiment Station (WES) was found to be the most promising.[46]

This model was based on Roesner and Norton's work with dissolved gas levels downstream of a spillway.[47] was developed by Roesner and Norton, who began with a simple mass transfer model that can be expressed as follows:

$$C_d = C_s - (C_s - C_u) e^{-Kt} \tag{12.1}$$

where ***Cd*** is the downstream concentration, ***Cs*** is the saturation concentration, ***Cu*** is the upstream concentration, ***K*** is the mass transfer coefficient, and ***t*** is the residence time in the stilling basin. This equation forms the basis of the WES model. Hibbs and Gulliver utilized this equation in computing the effective saturation concentration, ***Ce***:[48]

$$C_e = C_s \left(1 + \frac{d_e \gamma}{P_a} \right) \tag{12.2}$$

where ***Cs*** is the saturation concentration (taken to be 100%), ***de*** is the effective bubble depth, γ is the specific weight of water, and ***Pa*** is the atmospheric pressure. Geldert et al. provide three field data sets: Ice Harbor, The Dalles, and Little Goose. The measured effective saturation concentrations and the values computed using Equation 12.2 are shown in this next figure:

---

[45] Benton, D. J., "Benton, "Modeling of Nitrogen Supersaturation at Jennings Randolph," Advanced Technology Systems Report for the USACE, September, 1999. This report was part of the larger Section 1135(b) Study for Jennings Randolph Lake conducted under the Water Resources Development Act (WRDA) of 1986.

[46] Geldert, D. A., J. S. Gulliver, and S. C. Wilhelms (1998), "Modeling Dissolved Gas Supersaturation Below Spillway Plunge Pools," *Journal of Hydraulic Engineering*, May 1998, pp. 513-521.

[47] Roesner, L. A., and W. R. Norton, "A Nitrogen Gas (N_2) Model for the Lower Columbia River," Report No. 1-350, *Water Resources*, 1971.

[48] Hibbs, D. E. and J. S. Gulliver, "Prediction of an Effective Saturation Concentration at Spillway Plunge Pools," *Journal of Hydraulic Engineering*, Vol. 1, No. 3, pp. 940 949, 1997.

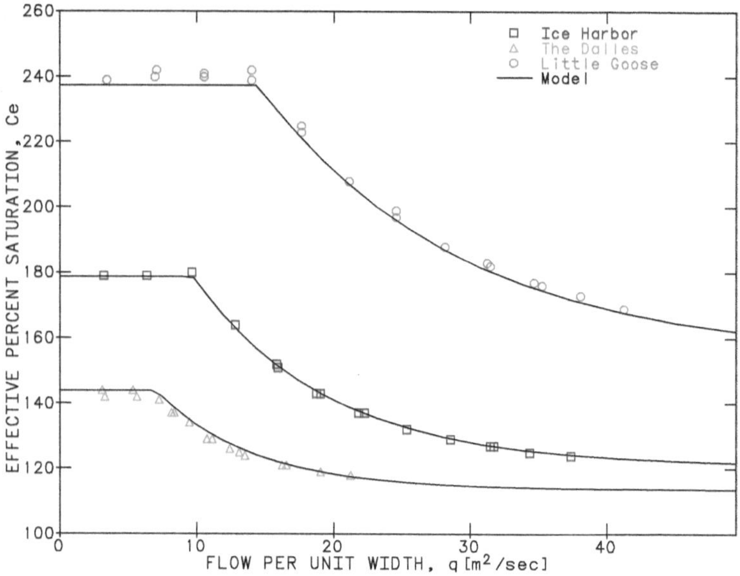

The effective depth is computed from the bubble half-life depth (i.e., the length traveled over the half-life), ***hb***, by Equation 12.3:

$$d_e = h_2 + (h_1 - h_2)\,e^{\left(1 - \frac{\beta h_b}{L_s}\right)} \quad for \quad \frac{\beta h_b}{L_s} > 1$$
$$d_e = h_1 \quad for \quad \frac{\beta h_b}{L_s} \leq 1 \qquad (12.3)$$

where $\boldsymbol{\beta}$ is an empirical constant equal to 2.2, $\boldsymbol{h_1}$ is the effective bubble depth in the stilling basin (presumed to be 2/3 of the stilling basin depth, ***hs***), $\boldsymbol{h_2}$ is the effective bubble depth in the river (presumed to be 1/2 of the river depth, ***hr***), and ***Ls*** is the length of the stilling basin. The bubble half-life depth is computed from the discharge per unit width, ***q***, and the bubble rise velocity, ***vr*** (presumed be constant at 0.25 meters/second), by Equation 12.4.

$$h_b = \frac{q}{v_r}\ln(2) \qquad (12.4)$$

Geldert et al. reasoned that the mass transfer included a bubble component into the water and a surface component out of the water. The rate of change of the concentration, ***C***, is then given by Equation 12.5:

$$\frac{dC}{dt} = K_L a_b (C_e - C) + K_L a_s (C_s - C) \qquad (12.5)$$

where $\boldsymbol{K_L}$ is the mass transfer coefficient, $\boldsymbol{a_b}$ is the bubble interfacial area per unit volume and $\boldsymbol{a_s}$ is the surface interfacial area per unit volume. The solution of this differential equation is given by Equation 12.6.

$$C_d = C_e - (C_e - C_u)\left\{ e^{-(K_L a_b t_b + K_L a_s t_s)} + \vartheta \right\}$$
$$\vartheta = \frac{K_L a_s t_s}{K_L a_b t_b + K_L a_s t_s}\left(\frac{C_e - C_s}{C_e - C_u}\right)\left[1 - e^{-(K_L a_b t_b + K_L a_s t_s)}\right] \tag{12.6}$$

where t_b is the residence time for the bubbles and ts is the exposure time for the surface transfer. Geldert et al. presumed that the combination $K_L a_s t_s$ would be a dimensionless constant on the order of unity for any particular application. The void fraction, φ, is computed using Equation 12.7:

$$\phi = \frac{v_j \lambda}{v_j \lambda + q} \tag{12.7}$$

where λ is an empirical constant on the order of 0.2 meters and v_j is the effective velocity of the plunging jet of water. Geldert et al. did not provide a means of obtaining v_j, simply stating that this was "computed by a standard water surface profile technique." Geldert et al. used the void fraction and an empirical correlation to obtain the dimensionless bubble transfer group, $K_L a_b t_b$, given by Equation 12.8:

$$K_L a_b t_b = \alpha\phi \frac{(1-\phi)^{1/2}}{\left(1-\phi^{5/3}\right)^{1/4}} W_e^{3/5} R_q^{2/3} S_c^{-1/2} R_r^{-1} \tag{12.8}$$

where α is an empirical constant on the order of unity, We is the Weber number (Equation 12.9), Rq is the Reynolds number for the flow (Equation 12.10), Sc is the Schmidt number for air/water (Equation 12.11), and Rr is the Reynolds number for the rising bubbles (Equation 12.12).

$$W_e = \frac{\rho q^2}{\sigma d_j} \tag{12.9}$$

where ρ is the density of water, σ is the surface tension, and d_j is the effective depth of the plunging jet ($d_j = q/v_j$).

$$R_q = \frac{q}{\upsilon} \tag{12.10}$$

where υ is the kinematic viscosity of water.

$$S_c = \frac{\nu}{D} \tag{12.11}$$

where D is the air/water diffusion coefficient.

$$R_r = \frac{2 d_e v_r}{\upsilon} \tag{12.12}$$

These equations form the original WES model. Modifications are required in order to complete the model and to obtain good agreement with field data. The original WES model also lacks an explicit calculation for the plunging jet velocity, v_j. In order to fill this gap in the model, a computer program was developed to "back out" the jet velocity implied by the data points for the three sites given in the WES report. These values were then compared to all of the dimensionless quantities that can be formed from the site parameters. The best correlation obtained (R^2=0.92) is given in Equation 12.13.

$$\frac{v_j}{\sqrt{g\,h_t}} = 0.15\left(\frac{\frac{q}{h_t}}{\sqrt{g\,h_t}}\right)^{0.23} \tag{12.13}$$

where g is the gravitational acceleration and ht is the total (effective) head. The agreement between the data and this equation is illustrated in the following figure.

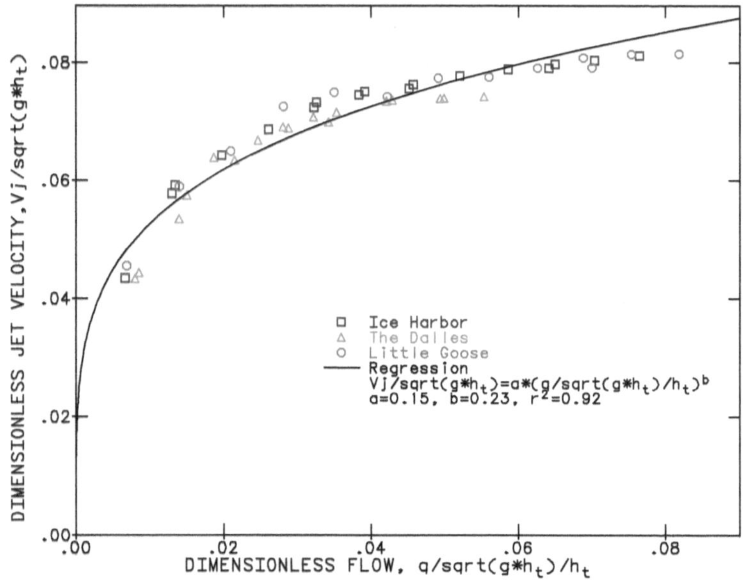

The bubble transfer group, $K_L a_b t_b$, can then be computed from Equation 12.8 and this correlation for the jet velocity (Equation 12.13). The results are illustrated in this next figure:

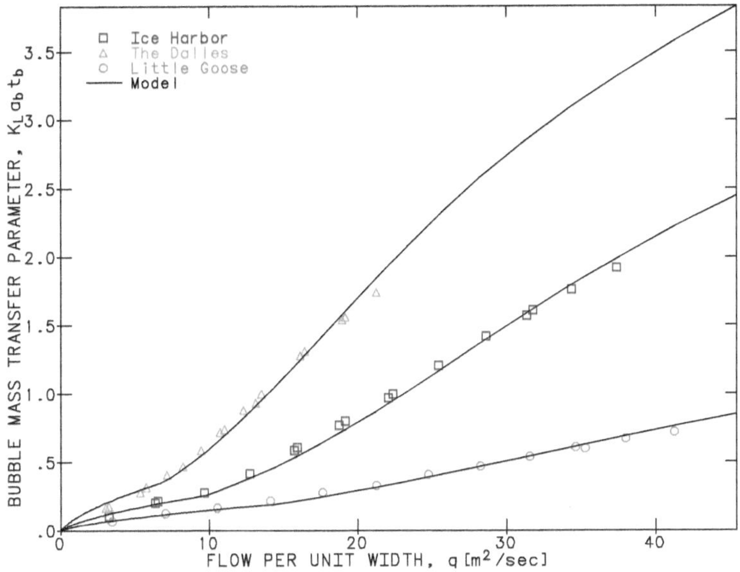

This same model can be applied to the Jennings Randolph site. The data and model results for all four sites are illustrated in the following figure:

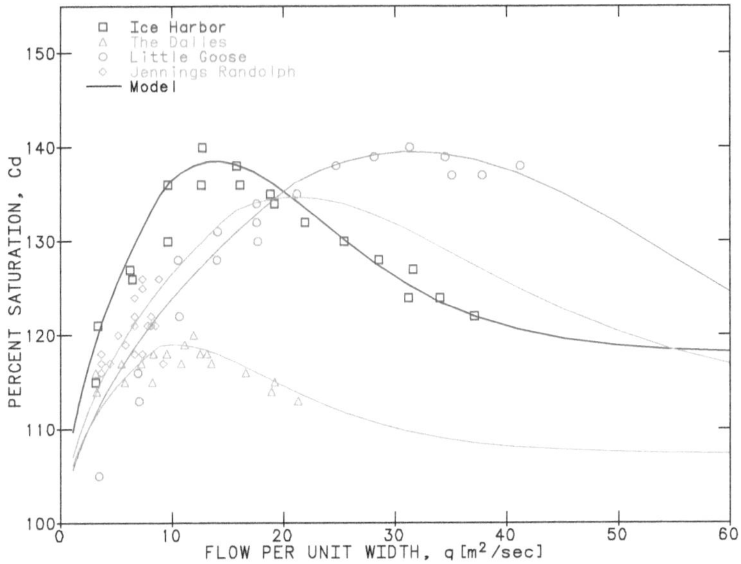

As stated previously, this is a zero-dimensional empirical model. The Modified WES Model is significantly different than a one-, two-, or three-dimensional finite difference or finite element model in which the domain is subdivided into computational cells. Any direct applications of this model are limited to the variables that appear in the various equations, for instance, a single value must represent the depth of the stilling basin, ***hs***. If the depth of the stilling basin changes significantly over its length, this model will only accommodate a single number for the average or effective depth. The source code implementing this model is provided in Appendix H.

Appendix A. Symbols & Terms

approach: the difference between the cold water (leaving) temperature and the ambient wet-bulb (for a cooling tower)

area per unit volume: It is not practical to measure the interfacial contact area between air and water droplets or sheets in cooling tower fill, but this is needed for mass transfer calculations. The fill (or packing) volume is easily measured. This parameter always appears multiplied by the mass or sensible heat transfer coefficients, so that the actual value is never needed.

backpressure: When a steam turbine exhausts into a condenser, the pressure in the condenser is "felt" at the turbine exit (i.e., "pushes" back).

dew point: the temperature at which condensation begins to form

drift eliminators: devices (usually in the shape of chevrons) designed to remove water droplets from moving air, thus reducing liquid carry-over and possibly pressure drop

dry-bulb: the conventional ambient air temperature

equilibrium temperature: For a cooling pond this is the temperature the pond would eventually reach, if all of the influences were held constant. The intent of this is to account for ambient temperature, ambient relative humidity, wind speed, and solar heating. Of course, this isn't possible, if for no other reason, the Sun doesn't remain at the same zenith.

rain zone: in a cooling tower beneath the fill (or packing) where the water droplets fall through the incoming air

range: the difference between the hot water (entering) temperature and the cold water (leaving) temperature (for a cooling tower)

wet-bulb: roughly equivalent to the adiabatic (no heat transfer) saturation temperature; measured by a temperature instrument covered with a wetted wick

Appendix B: Moist Air Property Functions

Accurate and fast moist air property functions are essential to all calculations involving evaporative cooling. The following VBA code provides all of the thermodynamic properties of moist air you will need. This code may also be found in several of the spreadsheets contained in the on-line archive that accompanies this text.

```
Option Explicit
'These properties are based on the 1993 ASHRAE Handbook
   of Fundamentals
'Chapter 6 Table 2 which is derived from the work of
   Hyland & Wexler
'The temperature and pressure dependence of the
   enhancement factor, f, are from
'Table 4 of this Section as well as Table 2 of Section 5
   of the 1977 Handbook.
'!!!!!!!!!!!!!!!!!!!!!!!!!!!!!!!!!!!!!!!!!!!!!!!!!!
'Note: relative humidity is 0 and 1, not 0 to 100!
'!!!!!!!!!!!!!!!!!!!!!!!!!!!!!!!!!!!!!!!!!!!!!!!!!!
Function fPs(Ts As Double) As Double
'saturation pressure of water
  Dim T As Double
  T = Ts + 459.67
  If (Ts < 32#) Then
    fPs = Exp((((-9.0344688E-14 * T + 3.5575832E-10) * T
   + 0.00000019202377) * T _
    - 0.0053765794) * T - 4.8932428 - 10214.165 / T +
   4.1635019 * Log(T))
  Else
    fPs = Exp(((-2.4780681E-09 * T + 0.00001289036) * T
   - 0.027022355) * T _
      - 11.29465 - 10440.397 / T + 6.5459673 * Log(T))
  End If
End Function
Function fT(T As Double) As Double
'temperaure dependence of the enhancement factor, f
  fT = ((((-1.22884575824E-13 * T + 2.98501566007E-11) *
   T - 1.68742180749E-09) * T _
    + 1.64960460192E-07) * T - 1.37450341076E-05) * T +
   1.00432811563436
End Function
Function fP(T As Double, P As Double) As Double
'pressure dependence of the enhancement factor
  fP = 0.9999278133 + (0.00002200673657 + (6.729181914E-
   08 - 7.590092091E-10 * T) * T) * T _
     + (0.0002997989681 + (-0.000001718254802 +
   8.054776534E-09 * T) * T - 3.40732948E-09 * T * P) *
   P
End Function
```

```
Function fTP(T As Double, P As Double) As Double
'composite (both temperature and pressure) dependence of
    the enhancement factor
  fTP = fT(T) * fP(T, P) / fP(T, 14.696)
End Function
Function fWdrh(Pbaro As Double, Tdb As Double, RH As
    Double) As Double
'humidity ratio from the pressure, dry-bulb, and
    relative humidity
  Dim f As Double, Ps As Double, Pw As Double
  f = fTP(Tdb, Pbaro)
  Ps = fPs(Tdb)
  Pw = f * RH * Ps
  fWdrh = 0.62198 * Pw / (Pbaro - Pw)
End Function
Function fPw(Pbaro As Double, W As Double) As Double
'partial pressure of water vapor in air, Pw=f*RH*Ps
  fPw = W * Pbaro / (0.62198 + W)
End Function
Function fDew(Pw As Double) As Double
'dew point from partial pressure
  Dim iter As Integer, T1 As Double, T2 As Double
  T1 = -80
  T2 = 200
  For iter = 1 To 32
    fDew = (T1 + T2) / 2
    If (fT(fDew) * fPs(fDew) < Pw) Then
      T1 = fDew
    Else
      T2 = fDew
    End If
  Next iter
End Function
Function fTdprh(Pbaro As Double, Tdb As Double, RH As
    Double) As Double
'dew point from dry-bulb and relative humidity
  fTdprh = fDew(fPw(Pbaro, fWdrh(Pbaro, Tdb, RH)))
End Function
Function fWdwb(Pbaro As Double, Tdb As Double, Twb As
    Double) As Double
'humidity ratio from pressure, dry-bulb, and wet-bulb
  Dim Ws As Double
  Ws = fWdrh(Pbaro, Twb, 1#)
  fWdwb = ((1093 - 0.556 * Twb) * Ws - 0.24 * Tdb + 0.24
    * Twb) / (1093 + 0.444 * Tdb - Twb)
End Function
Function fTdpwb(Pbaro As Double, Tdb As Double, Twb As
    Double) As Double
'dew point from dry-bulb and wet-bulb
```

```
  fTdpwb = fDew(fPw(Pbaro, fWdwb(Pbaro, Tdb, Twb)))
End Function
Function fTwdrh(Pbaro As Double, Tdb As Double, RH As
    Double) As Double
'wet-bulb from dry-bulb and relative humidity
  Dim iter As Integer, T1 As Double, T2 As Double, W As
   Double
  W = fWdrh(Pbaro, Tdb, RH)
  T1 = ((4.94704224634E-06 * Tdb - 0.00303815378496) *
   Tdb + 0.85025907521) * Tdb - 2.85417545122
  T2 = Tdb
  For iter = 1 To 32
    fTwdrh = (T1 + T2) / 2
    If (fWdwb(Pbaro, Tdb, fTwdrh) < W) Then
      T1 = fTwdrh
    Else
      T2 = fTwdrh
    End If
  Next iter
End Function
Function fRHdwb(Pbaro As Double, Tdb As Double, Twb As
    Double) As Double
'relative humidity from dry-bulb and wet-bulb
  Dim iter As Integer, R1 As Double, R2 As Double, W As
   Double
  W = fWdwb(Pbaro, Tdb, Twb)
  R1 = 0
  R2 = 1
  For iter = 1 To 32
    fRHdwb = (R1 + R2) / 2
    If (fWdrh(Pbaro, Tdb, fRHdwb) < W) Then
      R1 = fRHdwb
    Else
      R2 = fRHdwb
    End If
  Next iter
End Function
Function fTwdp(Pbaro As Double, Tdb As Double, Tdp As
    Double) As Double
'wet-bulb from dry-bulb and dew point
  Dim iter As Integer, T1 As Double, T2 As Double, W As
   Double
  W = fWdrh(Pbaro, Tdp, 1)
  T1 = ((4.94704224634E-06 * Tdb - 0.00303815378496) *
   Tdb + 0.85025907521) * Tdb - 2.85417545122
  T2 = Tdb
  For iter = 1 To 32
    fTwdp = (T1 + T2) / 2
    If (fWdwb(Pbaro, Tdb, fTwdp) < W) Then
```

```
      T1 = fTwdp
    Else
      T2 = fTwdp
    End If
  Next iter
End Function
Function fHdbw(Tdb As Double, W As Double) As Double
'enthalpy of moist air per pound of DRY air
  fHdbw = 0.24 * Tdb + W * (1061 + 0.444 * Tdb)
End Function
Function fHdrh(Pbaro As Double, Tdb As Double, RH As
    Double) As Double
'enthalpy of moist air per pound of DRY air
  fHdrh = fHdbw(Tdb, fWdrh(Pbaro, Tdb, RH))
End Function
Function fSdbw(Pbaro As Double, Tdb As Double, W As
    Double) As Double
'entropy of moist air per pound of DRY air
  Dim R As Double, Sg As Double
  R = 0.068686
  Sg = ((-4.33393735904E-08 * Tdb + 1.80117532949E-05) *
   Tdb - 0.00494503299302) * Tdb + 2.44380828045
  fSdbw = 0.24 * Log((Tdb + 459.67) / 459.67) + W * Sg -
   R * Log(Pbaro / 14.696)
End Function
Function fSdrh(Pbaro As Double, Tdb As Double, RH As
    Double) As Double
'enthalpy of moist air per pound of DRY air
  fSdrh = fSdbw(Pbaro, Tdb, fWdrh(Pbaro, Tdb, RH))
End Function
```

Appendix C. Runge-Kutta 2D for Crossflow Calculations

The 4th order Runge-Kutta method for solving two-dimensional partial differential equations (PDEs) is as follows:

$$k_1 = f(t_n, x_n, y_n)$$

$$l_1 = g(t_n, x_n, y_n)$$

$$k_2 = f(t_n + \tfrac{1}{2}h, x_n + \tfrac{1}{2}h\,k_1, y_n + \tfrac{1}{2}h\,l_1)$$

$$l_2 = g(t_n + \tfrac{1}{2}h, x_n + \tfrac{1}{2}h\,k_1, y_n + \tfrac{1}{2}h\,l_1)$$

$$k_3 = f(t_n + \tfrac{1}{2}h, x_n + \tfrac{1}{2}h\,k_2, y_n + \tfrac{1}{2}h\,l_2)$$

$$l_3 = g(t_n + \tfrac{1}{2}h, x_n + \tfrac{1}{2}h\,k_2, y_n + \tfrac{1}{2}h\,l_2)$$

$$k_4 = f(t_n + h, x_n + h\,k_3, y_n + h\,l_3)$$

$$l_4 = f(t_n + h, x_n + h\,k_3, y_n + h\,l_3)$$

$$k = \tfrac{1}{6}(k_1 + 2k_2 + 2k_3 + k_4),$$

$$l = \tfrac{1}{6}(l_1 + 2l_2 + 2l_3 + l_4)$$

$$x_{n+1} = x_n + h\,k$$

$$y_{n+1} = y_n + h\,l$$

$$t_{n+1} = t_n + h$$

The following code implements the method:

```
typedef void (*PDE)(double X,double Y,double
   U,double*dU,double V,double*dV);

void RungeKutta2D(PDE pde,double X,double dX,double
   Y,double dY,double*U,double*V)
  {
  double dU,dU1,dU2,dU3,dU4,dV,dV1,dV2,dV3,dV4;
  pde(X,Y,U[0],&dU1,V[0],&dV1);
  pde(X+dX/2.,Y+dY/2,U[0]+dU1*dX/2,&dU2,
   V[0]+dV1*dY/2,&dV2);
  pde(X+dX/2.,Y+dY/2.,U[0]+dU2*dX/2.,&dU3,
   V[0]+dV2*dY/2.,&dV3);
  pde(X+dX,Y+dY,U[0]+dU3*dX,&dU4,V[0]+dV3*dY,&dV4);
  dU=(dU1+2.*dU2+2.*dU3+dU4)/6.;
  dV=(dV1+2.*dV2+2.*dV3+dV4)/6.;
  U[0]+=dU*dX;
  V[0]+=dV*dY;
  }
```

The PDE is supplied by a function, for example:

```
void Cell(double X,double Y,double Ha,double*dHa,double
   Tw,double*dTw)
```

```
{
double Hw,Q;
Hw=fHtwb(Pbaro,Tw);
Q=KaY*(Hw-Ha);
dHa[0]=Q*LG;
dTw[0]=-Q;
}
```

The process of stepping through the domain is illustrated in the following code:

```
for(y=0;y<Ny;y++)
  {
  for(x=0;x<Nx;x++)
    {
    H=Ha[(Nx+1)*y+x];
    T=Tw[Nx*y+x];
    RungeKutta2D(Cell,X,1./Nx,Y,1./Ny,&H,&T);
    Ha[(Nx+1)*y+x+1]=H;
    Tw[Nx*(y+1)+x]=T;
    }
  }
```

Appendix D. Falling Droplet Trajectory and Mass Transfer

The 4th order Runge-Kutta method for solving one-dimensional ordinary differential equations (ODEs) is as follows:

```
void RungKutta4( /* one step of 4th order Runge-Kutta
    integration */
  void dydx(double,double*,double*), /* function
    returning dY/dX */
  double*x, /* independent variable */
  double dx, /* step in X */
  double*y, /* dependent variable array */
  double*dy, /* step in Y array */
  int n) /* number of dependent variables */
  {
  int i,j;
  double a[4]={0,.5,.5,1};
  double b[4]={1/6.,1/3.1/3.,1/6.};
  double*v; /* working array of dimension n */
  double*w; /* working array of dimension 4n */
  w=calloc(4*n,sizeof(double));
  v=calloc(  n,sizeof(double));
  dydx(x[0],y,w);
  for(j=1;j<4;j++)
    {
    for(i=0;i<n;i++)
      {
      dy[i]=a[j]*w[n*(j-1)+i];
      v[i]=y[i]+dx*dy[i];
      }
    dydx(x[0]+dx*a[j],v,w+n*j);
    }
  for(i=0;i<n;i++)
    {
    dy[i]=0;
    for(j=0;j<4;j++)
      dy[i]+=b[j]*w[n*j+i];
    y[i]+=dx*dy[i];
    }
  x[0]+=dx;
  free(w);
  free(v);
  }
```

The function providing the derivatives is listed below:

```
void DropTrajectory(double t,double*Q,double*dQ) /*
    trajectory of a droplet falling */
  { /* in a moving stream of air */
  double As; /* droplet surface area [ft²] */
```

```
double Av; /* drop surface area per fill volume [1/ft]
 */
double Ax; /* droplet cross-sectional area [ft²] */
double Cd; /* drag coefficient */
double d; /* drop diameter [ft] */
double Dc; /* diffusion coefficient [ft²/sec] */
double Dv; /* drop volume [ft^3] */
double Nu; /* Nusselt number */
double Re; /* Reynolds number */
double Sh; /* Sherwood number */
double Ud; /* horizontal drop velocity component
 [ft/sec] */
double Ur; /* relative horizontal velocity component
 [ft/sec] */
double Vd; /* vertical drop velocity component
 [ft/sec] */
double Vr; /* relative vertical velocity component
 [ft/sec] */
double Ws; /* absolute humidity at saturation */
double y; /* net vertical displacement of drop [ft] */
d=max(1E-10,pow(6*Q[7]/Rw/pi,1/3.));/* compute drop
 diameter from mass */
As=pi*d*d; /* drop surface area */
Ax=As/4; /* drop cross-sectional area */
Dv=As*d/6; /* drop volume */
y=max(.000001,-Q[1]); /* net vertical movement */
tH=t*Hdrop/y; /* compute hold-up time */
Td=32+Q[6]/Dv/Rw/Cp;/* compute drop temperature from
 energy */
Ws=fWs(Td); /* absolute humidity at saturation */
Ud=Q[2]; /* absolute horizontal velocity component */
Vd=Q[3]; /* absolute vertical velocity component */
Ur=Ua-Ud; /* relative horizontal velocity component */
Vr=Va-Vd; /* relative vertical velocity component */
Re=sqrt(Ur*Ur+Vr*Vr)*d*Ra/mU; /* Reynolds number */
Cd=Drag(Re); /* drag coefficient */
dP=(Q[4]*Ua/Vo+Q[5]*Va/Vo)*Fd/y/(Ra*Vo*Vo/2/g);/*
 pressure drop */
Dc=mU/Ra/Sc; /* diffusion coefficient */
Nu=Nusselt(Re,Pr); /* Nusselt number */
Sh=Sherwood(Re,Sc); /* Sherwood number */
Av=As*Fd*t/y; /* drop surface area per unit of fill
 volume */
KaYL=y*Sh*Ra*Dc/d*Av/(Fl/3600); /* mass transfer
 parameter (KaY/L) */
dQ[0]=Q[2]; /* diff horizontal position = horizontal
 velocity */
dQ[1]=Q[3]; /* diff vertical position = vertical
 velocity */
```

```
dQ[2]=3*Cd*Ra*Ur*fabs(Ur)/4/d/Rw; /* diff horizontal
  velocity from horizontal impulse */
dQ[3]=(3*Cd*Ra*Vr*fabs(Vr)-4*d*(Rw-Ra)*g)/4/Rw/d; /*
  diff vertical velocity from vertical impulse */
dQ[4]=Cd*Ra*Ur*fabs(Ur)*Ax/2/g; /* diff horizontal
  impulse from horizontal force balance */
dQ[5]=Cd*Ra*Vr*fabs(Vr)*Ax/2/g; /* diff vertical
  impulse from vertical force balance */
dQ[6]=(Tdb-Td)*Nu*Ka*As/d+Hfg*Ra*(Wa-Ws)*Sh*Dc*As/d;/*
  diff energy from heat and mass transfer */
dQ[7]=Ra*(Wa-Ws)*Sh*Dc*As/d; /* diff mass from
  evaporation */
}
```

The Nusselt number, Sherwood number, and drag coefficient are provided by the following code:

```
double Drag(double Re) /* drag coefficient for a sphere
   */
  {
  return(0.22+(1+0.15*pow(Re,0.6))*24/Re);
  }

double Nusselt(double Re,double Pr) /* Nusselt number
   for a sphere */
  { /* from Kreith page 473 */
  return(2+(.4*pow(Re,.5)+.06*pow(Re,.67))*pow(Pr,.4));
  }

double Sherwood(double Re,double Sc) /* Sherwood number
   for a sphere */
  { /* from Treybal page 68 */
  return(2+(.4*pow(Re,.5)+.06*pow(Re,.67))*pow(Sc,.4));
  }
```

The solution process is carried out by the following code:

```
int Droplet(double Do,double fL,double hD,double
   vO,double angle,double To,double tDB,double rh)
  {
  double As,d,dQ[8],dt,Dv,Q[8],t;
  Fl=fL;
  Hdrop=hD;
  Vo=vO;
  Tdb=tDB;
  approach=range=0;
  t=0; /* initial time */
  dt=.01; /* time step */
  Td=To; /* initial drop temperature */
  Do*=FeetPerMm; /* convert drop diameter to ft */
  As=pi*Do*Do; /* droplet surface area */
  Dv=As*Do/6; /* droplet volume */
```

```
Fd=Fl/Rw/3600/Dv; /* droplet mass flux */
memset(Q,0,8*sizeof(double));
Q[6]=Dv*Rw*Cp*(To-32); /* initial energy of droplet */
Q[7]=Dv*Rw; /* initial mass of droplet */
Wa=fWdbr(Tdb,rh); /* ambient absolute humidity */
Twb=fBdbr(Tdb,rh); /* ambient wet-bulb */
Ra=fDdbw(Po,Tdb,Wa); /* ambient air density */
Ua=Vo*cos(angle*pi/180); /* horizontal air velocity
  component */
Va=Vo*sin(angle*pi/180); /* vertical air velocity
  component */
while(-Q[1]<Hdrop) /* step until droplet falls Hd */
   RungKutta4(DropTrajectory,&t,dt,Q,dQ,8);
approach=Td-Twb;
range=To-Td;
return(Td>Twb+.5);
}
```

The following is typical output from this program:

```
DROPS/V3.10: water droplets falling in flowing air
             by Dudley J. Benton, Knoxville, Tennessee

Do ..... initial drop diameter [mm] (1 to 10)
L ...... mass flux of water droplets [lbm/hr/sq.ft] (100 to
100000)
H ...... vertical distance of drop fall [ft] (1 to 10)
Va ..... air velocity [ft/sec] (1 to 100)
angle .. angle of inclination of airflow [degrees] (-180 to
180)
Tid .... initial temperature of drop [F] (32 to 150)
Tdb .... dry-bulb temperature [F] (-20 to 120)
RH ..... relative humidity [%] (0 to 100)

enter Do,L,H,Va,angle,Tid,Tdb,RH 3 5000 30 10 0 90 60 50

L/G=1.826, Twb=50.2, Wa=0.0055, Ra=0.0761
 time     D      X      Y      U      V  KaY/L     dP  holdup Tdrop
 0.00  3.00    0.0    0.0    0.0    0.0 ***** ****** ******  90.0
 0.10  3.00    0.0   -0.0    0.2   -1.4 0.027  2.152 73.571  89.6
 0.20  3.00    0.0   -0.1    0.4   -2.8 0.055  1.131 41.858  89.2
 0.30  3.00    0.0   -0.3    0.6   -4.2 0.083  0.758 29.274  88.8
 0.40  3.00    0.1   -0.5    0.8   -5.6 0.114  0.564 22.553  88.5
 0.50  3.00    0.1   -0.8    1.0   -6.9 0.146  0.446 18.382  88.1
 0.60  3.00    0.2   -1.1    1.1   -8.2 0.180  0.367 15.549  87.7
 0.70  3.00    0.2   -1.5    1.3   -9.5 0.217  0.310 13.502  87.3
 0.80  3.00    0.3   -1.9    1.5  -10.7 0.255  0.267 11.958  86.9
 0.90  3.00    0.3   -2.4    1.6  -11.8 0.295  0.234 10.754  86.5
 1.00  3.00    0.4   -3.0    1.7  -12.9 0.337  0.208  9.790  86.1
 1.10  3.00    0.5   -3.6    1.9  -14.0 0.381  0.187  9.003  85.7
 1.20  3.00    0.6   -4.2    2.0  -15.0 0.425  0.169  8.349  85.3
 1.30  3.00    0.7   -4.9    2.1  -16.0 0.471  0.155  7.797  84.9
 1.40  3.00    0.8   -5.6    2.2  -16.9 0.518  0.142  7.327  84.4
 1.50  3.00    0.9   -6.4    2.3  -17.7 0.566  0.131  6.921  84.0
 1.60  2.99    1.0   -7.2    2.4  -18.6 0.614  0.122  6.569  83.6
```

1.70	2.99	1.1	-8.0	2.5	-19.3	0.663	0.114	6.260	83.2
1.80	2.99	1.2	-8.9	2.6	-20.0	0.713	0.107	5.987	82.8
1.90	2.99	1.3	-9.8	2.7	-20.7	0.762	0.101	5.745	82.4
2.00	2.99	1.4	-10.7	2.8	-21.3	0.812	0.095	5.529	81.9
2.10	2.99	1.6	-11.7	2.9	-21.9	0.863	0.090	5.335	81.5
2.20	2.99	1.7	-12.7	3.0	-22.4	0.913	0.086	5.160	81.1
2.30	2.99	1.8	-13.7	3.1	-23.0	0.964	0.082	5.002	80.7
2.40	2.99	2.0	-14.7	3.2	-23.4	1.015	0.078	4.859	80.3
2.50	2.99	2.1	-15.7	3.3	-23.9	1.065	0.075	4.728	79.9
2.60	2.99	2.3	-16.8	3.3	-24.3	1.116	0.072	4.608	79.5
2.70	2.99	2.4	-17.9	3.4	-24.6	1.166	0.069	4.498	79.1
2.80	2.99	2.6	-19.0	3.5	-25.0	1.217	0.067	4.397	78.7
2.90	2.99	2.7	-20.1	3.6	-25.3	1.267	0.064	4.304	78.4
3.00	2.99	2.9	-21.2	3.6	-25.6	1.317	0.062	4.218	78.0
3.10	2.99	3.0	-22.3	3.7	-25.9	1.368	0.060	4.138	77.6
3.20	2.99	3.2	-23.5	3.8	-26.1	1.418	0.058	4.064	77.2
3.30	2.99	3.4	-24.6	3.8	-26.3	1.467	0.056	3.995	76.9
3.40	2.99	3.5	-25.8	3.9	-26.6	1.517	0.055	3.931	76.5
3.50	2.99	3.7	-27.0	4.0	-26.8	1.567	0.053	3.871	76.2
3.60	2.99	3.9	-28.2	4.0	-26.9	1.616	0.052	3.815	75.8
3.70	2.99	4.1	-29.3	4.1	-27.1	1.665	0.050	3.763	75.5
3.76	2.99	4.2	-30.1	4.1	-27.2	1.694	0.050	3.733	75.3

range=14.7, approach=25.1

NOTE: drop temperature applies only to solitary drops; whereas, KaY/L, dP, and holdup apply to a population.

Appendix E. Nuclear Plant Thermal Performance

The following functions may be used to model the thermal performance of a large (c. 1000 MWe) power plant. These particular functions are for a Westinghouse pressurized water reactor. The condenser is in three sections, producing three sequentially higher backpressures. The pump curve (head vs. flow) and the hydraulic resistance is used to calculate the condenser cooling water (CCW) flow for one, two, and three pumps, which is not linear.

```
/*****************************************************
***********************
  APPROXIMATE STEAM PROPERTIES
*****************************************************
***********************/

double fTsat(double P)            /* Tsat of steam in øF
from pressure in psia */
  {
  double A;
  A=log(P);
  return((26.5029*A+101.692)/(-0.0683446*A+1.));
  }

double fPsat(double T)                /* Psat of steam in
psia from Tsat in øF */
  {
  return(exp((0.0377315*T-3.83698)/(0.00257868*T+1.)));
  }

/*****************************************************
***********************
  HEAT EXCHANGE INSTITUTE STANDARDS FOR STEAM SURACE
CONDENSERS 1989 ED.
*****************************************************
***********************/

/* standard tubing wall thickness for gages 12-24 */
double
Wall[13]={0.109,0.095,0.083,0.072,0.065,0.058,0.049,0.04
2,0.035,0.032,0.028,0.025,0.022};

double WaterTemperatureCorrectionFactor(double T)     /*
Heat Exch. Inst. 1989 */
  {                                              /*
empirical correction factor  */
  return(0.376769/((((2.15973E-9*T-1.13681E-
6)*T+0.000221828)*T-0.0195586)*T+1.));
  }
```

```
/*****************************************************
**********************
  GENERIC PIPE FLOW FUNCTIONS
*****************************************************
**********************/

double Colebrook(double Re,double k)                  /*
Colebrook's formula   */
  {                                                   /*
Re=Reynolds number    */
  double f;                                           /*
k=relative roughness  */
  int iter;                                           /*
f=friction factor     */
  if(Re<2300.)                                        /*
dP=f*(L/D)*rho*V²/2/g */
    return(64./Re);
  f=0.01;
  for(iter=0;iter<5;iter++)
    {
    f=1.14-2.*(log10(k)+log10(1.+9.3/Re/k/sqrt(f)));
    f=1./f/f;
    }
  if(Re<5500.)
    f=64./Re+(log10(Re)-3.3617)/0.3372*(f-64./Re);
  return(f);
  }

double WoodsF(double Re,double k)        /* Wood's
formula for friction factor */
  {                                      /* Re>1E4 see
Colebrook() for details */
  double a,b,c;
  a=0.094*pow(k,0.225)+0.53*k;
  b=88.*pow(k,0.44);
  c=1.62*pow(k,0.134);
  return(a+b/pow(Re,c));
  }

/*****************************************************
**********************
  PLANT-SPECIFIC THERMAL CALCULATIONS
*****************************************************
**********************/

int    gage       =   22; /* condenser tube gage */
int    Ntubes     =44764; /* number of tubes per
condenser */
```

```
double Dtube       =   1.; /* condenser tube outside
diameter [in] */
double Ltube       =91.73; /* total length of condenser
tubes [ft] */
double cleanliness=  90.; /* default condenser
cleanliness [%] */
double material    = 0.79; /* HEI empirical tube material
factor */

double HeatRateCorrectionFactor(double pl,double bp)
/* change in heatrate */
  {                                                  /*
pl=power level (1=100%)     */
  double Ts;                                         /*
bp=back pressure [in.Hg]    */
  Ts=fTsat(bp*0.4911540775);                         /*
Ts=steam saturation temp.   */
  return(((-0.0337506*pl+(0.762123*Ts-76.9946))*pl+
    ((-0.00484629*Ts-0.641417)*Ts+114.355))*pl
    +((2.75779E-5*Ts-0.000100014)*Ts+0.12553)*Ts-
40.1616);
  }

double GeneratorOutput(double rh,double bp)   /*
electrical power output [MWe] */
  {                                           /*
rh=reactor heat input [MWt]    */
  double gen,hrcf,locf,pl;                    /* bp=back
pressure [in.Hg]       */
  pl=rh/3425.;                                /* pl=power
level (1=100%)         */
  hrcf=HeatRateCorrectionFactor(pl,bp);       /*
hrcf=heatrate correction fact */
  locf=100./(100.+hrcf);                      /*
locf=load correction factor % */
  gen=1175.*((-0.10336*pl+1.256)*pl-.15264);/*
gen=uncorrected generator MWe */
  return(max(0,gen*locf));
  }

double CondenserHeatRejection(double rh,double bp) /*
cond. ht. rej. [BTU/hr] */
  {                                                /*
rh=reactor heat input [MWt] */
  double go,qr;                                    /*
bp=back pressure [in.Hg]    */
  go=GeneratorOutput(rh,bp);                       /*
go=generator output [MWe]   */
```

```
  qr=rh-go;                                         /*
qr=gross heat reject. [MWt] */
  return(0.98*qr*3412140.);                         /* about
98% goes to condenser */
  }

double CondenserRise(double rh,double bp,double gpm) /*
water temp. rise [øF] */
  {                                                 /*
rh=reactor heat input [MWt] */
  double qr;                                        /*
bp=back pressure [in.Hg]    */
  qr=CondenserHeatRejection(rh,bp);                 /*
gpm=cooling water flow      */
  return(qr/gpm/8.335/60.);                         /*
qr=cond. heat rejt [BTU/hr] */
  }

double PumpCurve(double gpm)              /* CCW pump head
[ft] from flow [gpm] */
  {
  double x;
  x=gpm/247000.;
  return(45.*((((1.64822*x-6.75411)*x+8.2608)*x-
4.28862)*x+2.13944));
  }

double CondenserFlow(                                /*
condenser flow [gpm]      */
  int    pumps,                                      /*
number of pumps operating */
  int    tubes,                                      /*
number of active tubes    */
  double Di,                                         /* tube
inside diameter [in] */
  double length)                                     /* tube
length [ft]          */
  {
  int iter;
  double
area,drop,friction,gpm,gpm1,gpm2,head,Reynolds,velocity;

  if(pumps<1||tubes<10000||Di<0.5||length<10)     /*
check for null conditions */
    return(0);

  Di/=12.;                                            /*
convert diameter to feet */
```

```
  area=tubes*M_PI*Di*Di/4.;                                  /*
inside area of tubes */
  gpm=pumps*247000.;                                     /*
initial estimate of flow */

  gpm1=1000.;
/* lower bound on flow */
  gpm2=2.*gpm;
/* upper bound on flow */

  for(iter=0;iter<32;iter++)                                 /*
use bisection search */
    {                                                        /*
to match drop & head */
    gpm=(gpm1+gpm2)/2.;
    velocity=gpm/area/7.48052/60.;                        /*
water velocity [ft/sec] */
    Reynolds=velocity*Di/0.0000122;
/* Reynolds number */
    friction=WoodsF(Reynolds,0.00006);
/* friction factor */

    drop=12.
/* static head loss */
        +12.16569*pow(gpm/729350.,2)                    /*
waterbox + etc. head loss */
        +friction*(length/Di)*                            /*
tube friction head loss */
          velocity*velocity/2./32.174;
    head=PumpCurve(gpm/pumps);                             /*
pump head at this flow */

    if(drop<head)
/* adjust flow so that */
      gpm1=gpm;
/* drop and head match */
    else
      gpm2=gpm;
    }

  return(gpm);
  }

double BackpresGeneratorOutput(
  /*** function input parameters
***********************************************/
  double  Wt,    /* reactor heat input [MWt]   *   These
calculations are from   */
```

```
  double  gpm,    /* condenser flow [gpm]       *   the
Heat Exchange Institute    */
  double  length,/* tube length [ft]            *
Standards for Steam Surface   */
  int tubes,      /* number of active tubes     *
Condensers, 1989 Edition.     */
  double  Cc,     /* cleanliness factor [%]
**********************************/
  double  Cm,     /* empirical material factor  */
  double  Do,     /* outside tube diameter [in] */
  double  Di,     /* inside tube diameter [in]  */
  double  Ti,     /* inlet water temp [øF]      */
  /*** function output parameters ***********/
  double  bp[3], /* zone backpressures [in.Hg] */
  double  Te[3]) /* exit water temps [øF]      */
  {
  int iter,zone;
  double
area,bp1,bp2,Ct,Cv,effectiveness,excess,mwe,rise,Th,Tsat
,NTU,velocity;

  if(Wt<856.||gpm<100000.)
    {
    Te[0]=Te[1]=Te[2]=Ti;
    bp[0]=bp[1]=bp[2]=0.;
    return(0.);
    }

  Ct=0.107*Do/Di/Di;                  /* empirical tube
diameter and gage factor */

  Di/=12.;                                          /*
convert diameter to feet */
  area=Ntubes*M_PI*Di*Di/4.;
/* flow area [ft²] */

  velocity=gpm/area/7.48052/60.;                     /*
water velocity [ft/sec] */

  Cv=263./sqrt(velocity);                       /* HEI
empirical velocity factor */

  NTU=Ct*Cv*Cm*(Cc/100.)                        /* number
of heat transfer units */
     *(length/3.)/7.48052/60.;               /* (simplified
heat exchange model) */

  effectiveness=exp(-
NTU*WaterTemperatureCorrectionFactor(Ti));
```

```
  excess=effectiveness/(1-effectiveness);

  bp[0]=bp[1]=bp[2]=fPsat(Ti)/0.4911540775;   /* initial
estimate of backpres */

  for(iter=0;iter<32;iter++)                          /*
iterate to converge on */
    {
/* zone backpressures */
    for(zone=0,Th=Ti;zone<3;zone++)
      {
      rise=CondenserRise(Wt,bp[zone],gpm)/3.;
      Th+=rise;
      Te[zone]=Th;
      Tsat=Th+rise*excess;
      bp1=bp[zone];
      bp2=fPsat(Tsat)/0.4911540775;
      bp[zone]=sqrt(bp1*bp2);                  /* use sqrt()
to dampen iterations */
      }
    if(iter>2&&fabs(bp1-bp2)<0.001)                    /*
check for convergence */
      break;
    }

  for(mwe=zone=0;zone<3;zone++)                       /*
compute generator output */
    mwe+=GeneratorOutput(Wt,bp[zone])/3.;
  return(mwe);
  }

void PlantTables()
  {
  int pumps;
  double bp[3],gpm,mwe,mwt,pl,qrej,Te[3],Ti;

  printf("Typical Nuclear Power Plant Performance\n");

  printf("\n CCW Flow\n");
  printf("pumps  gpm\n");
  for(pumps=0;pumps<=3;pumps++)
    {
    gpm=CondenserFlow(pumps,Ntubes,Dtube-2.*Wall[gage-
12],Ltube);
    printf("  %i %6.0lf\n",pumps,gpm);
    }

  printf("\n    Heat Rate Correction\n");
  printf(" bp   100%%    75%%    50%%    25%%\n");
```

```
  for(bp[2]=0.5;bp[2]<=5.5;bp[2]+=0.5)
    {
    printf("%3.1lf",bp[2]);
    for(pl=1.;pl>0.;pl-=0.25)
      printf("
%5.1lf%%",HeatRateCorrectionFactor(pl,bp[2]));
    printf("\n");
    }

  printf("\nPower & Heat Reject\n");
  printf(" MWt    MWe MBTU/hr\n");
  for(mwt=1000.;mwt<=3500.;mwt+=250.)
    printf("%4.0lf %6.1lf
%6.1lf\n",mwt,GeneratorOutput(mwt,2.),CondenserHeatRejec
tion(mwt,2.)/1E6);

  printf("\n    Generator Output [MWe]\n");
  printf(" bp   100%%    75%%    50%%    25%%\n");
  for(bp[2]=0.5;bp[2]<=5.5;bp[2]+=0.5)
    {
    printf("%3.1lf",bp[2]);
    for(pl=1.;pl>0.;pl-=0.25)
      {
      mwt=pl*3425.;
      printf(" %6.1lf",GeneratorOutput(mwt,bp[2]));
      }
    printf("\n");
    }

  printf("\nCondenser Heat Reject [MBTU/hr]\n");
  printf(" bp   100%%    75%%    50%%    25%%\n");
  for(bp[2]=0.5;bp[2]<=5.5;bp[2]+=0.5)
    {
    printf("%3.1lf",bp[2]);
    for(pl=1.;pl>0.;pl-=0.25)
      {
      mwt=pl*3425.;
      printf("
%6.1lf",CondenserHeatRejection(mwt,bp[2])/1E6);
      }
    printf("\n");
    }

  printf("\n  3 Pump Condenser Rise [øF]\n");
  printf(" bp   100%%    75%%    50%%    25%%\n");
  for(bp[2]=0.5;bp[2]<=5.5;bp[2]+=0.5)
    {
    printf("%3.1lf",bp[2]);
    for(pl=1.;pl>0.;pl-=0.25)
```

```
        {
        mwt=pl*3425.;
        printf(" %6.2lf",CondenserRise(mwt,bp[2],gpm));
        }
      printf("\n");
      }

    printf("\n  3 Pump Backpressure [in.Hg]\n");
    printf(" Ti   100%%    75%%    50%%    25%%\n");
    for(Ti=35.;Ti<=125.1;Ti+=5.)
      {
      printf("%3.0lf",Ti);
      for(pl=1.;pl>0.;pl-=0.25)
        {
        mwt=pl*3425.;

  mwe=BackpresGeneratorOutput(mwt,gpm,Ltube,Ntubes,cleanli
  ness,material,Dtube,Dtube-2.*Wall[gage-12],Ti,bp,Te);
        printf(" %6.3lf",bp[2]);
        }
      printf("\n");
      }

    printf("\n  3 Pump Generator Output [MWe]\n");
    printf(" Ti   100%%    75%%    50%%    25%%\n");
    for(Ti=35.;Ti<=125.1;Ti+=5.)
      {
      printf("%3.0lf",Ti);
      for(pl=1.;pl>0.;pl-=0.25)
        {
        mwt=pl*3425.;

  mwe=BackpresGeneratorOutput(mwt,gpm,Ltube,Ntubes,cleanli
  ness,material,Dtube,Dtube-2.*Wall[gage-12],Ti,bp,Te);
        printf(" %6.1lf",mwe);
        }
      printf("\n");
      }

    printf("\nCondenser Heat Reject [MBTU/hr]\n");
    printf(" Ti   100%%    75%%    50%%    25%%\n");
    for(Ti=35.;Ti<=125.1;Ti+=5.)
      {
      printf("%3.0lf",Ti);
      for(pl=1.;pl>0.;pl-=0.25)
        {
        mwt=pl*3425.;
```

```
mwe=BackpresGeneratorOutput(mwt,gpm,Ltube,Ntubes,cleanli
ness,material,Dtube,Dtube-2.*Wall[gage-12],Ti,bp,Te);
      qrej=60.*8.335*gpm*(Te[2]-Ti)/1E6;
      printf(" %6.1lf",qrej);
      }
    printf("\n");
    }

  printf("\n    Condenser Rise [øF]\n");
  printf(" Ti  100%%   75%%   50%%   25%%\n");
  for(Ti=35.;Ti<=125.1;Ti+=5.)
    {
    printf("%3.0lf",Ti);
    for(pl=1.;pl>0.;pl-=0.25)
      {
      mwt=pl*3425.;

mwe=BackpresGeneratorOutput(mwt,gpm,Ltube,Ntubes,cleanli
ness,material,Dtube,Dtube-2.*Wall[gage-12],Ti,bp,Te);
      printf(" %5.2lf",Te[2]-Ti);
      }
    printf("\n");
    }
  }
```

The function PlantTables() lists the calculated performance for a range of input variables and also illustrates how to call the functions. The output is as follows:

```
Typical Nuclear Power Plant Performance

 CCW Flow
pumps  gpm
  0      0
  1 317758
  2 569330
  3 733382

    Heat Rate Correction
 bp    100%     75%     50%     25%
0.5    0.1%   -0.7%   -5.6%  -14.6%
1.0   -0.7%   -1.7%   -4.8%  -10.0%
1.5   -0.6%   -1.2%   -2.6%   -5.0%
2.0   -0.0%    0.0%   -0.0%   -0.0%
2.5    0.9%    1.5%    2.8%    4.9%
3.0    2.0%    3.2%    5.8%    9.7%
3.5    3.3%    5.1%    8.8%   14.4%
4.0    4.6%    7.1%   11.9%   19.0%
4.5    6.1%    9.1%   14.9%   23.4%
5.0    7.6%   11.2%   18.0%   27.8%
5.5    9.1%   13.3%   21.0%   32.1%
```

Power & Heat Reject

MWt	MWe	MBTU/hr
1000	241.2	2537.4
1250	343.1	3032.6
1500	443.7	3532.2
1750	543.0	4036.1
2000	641.0	4544.3
2250	737.7	5056.9
2500	833.2	5573.8
2750	927.3	6095.0
3000	1020.1	6620.5
3250	1111.7	7150.3
3500	1201.9	7684.5

Generator Output [MWe]

bp	100%	75%	50%	25%
0.5	1173.3	865.5	559.7	213.0
1.0	1182.8	873.6	554.5	202.1
1.5	1181.9	869.2	542.4	191.6
2.0	1175.0	859.2	528.2	182.0
2.5	1164.5	846.4	513.6	173.5
3.0	1151.8	832.1	499.2	165.9
3.5	1137.8	817.3	485.3	159.1
4.0	1122.9	802.2	472.1	153.0
4.5	1107.7	787.3	459.6	147.5
5.0	1092.2	772.5	447.8	142.4
5.5	1076.7	758.1	436.6	137.8

Condenser Heat Reject [MBTU/hr]

bp	100%	75%	50%	25%
0.5	7529.5	5695.6	3854.9	2150.9
1.0	7497.8	5668.4	3872.1	2187.3
1.5	7500.7	5683.1	3912.7	2222.7
2.0	7523.8	5716.7	3960.2	2254.6
2.5	7558.8	5759.5	4009.1	2283.0
3.0	7601.3	5807.0	4057.2	2308.4
3.5	7648.2	5856.7	4103.5	2331.1
4.0	7697.9	5907.0	4147.6	2351.6
4.5	7749.0	5957.1	4189.5	2370.1
5.0	7800.8	6006.4	4229.1	2387.0
5.5	7852.6	6054.7	4266.6	2402.5

3 Pump Condenser Rise [øF]

bp	100%	75%	50%	25%
0.5	20.53	15.53	10.51	5.86
1.0	20.44	15.46	10.56	5.96
1.5	20.45	15.50	10.67	6.06
2.0	20.51	15.59	10.80	6.15
2.5	20.61	15.70	10.93	6.22
3.0	20.73	15.83	11.06	6.29
3.5	20.85	15.97	11.19	6.36
4.0	20.99	16.11	11.31	6.41
4.5	21.13	16.24	11.42	6.46
5.0	21.27	16.38	11.53	6.51
5.5	21.41	16.51	11.63	6.55

3 Pump Backpressure [in.Hg]

Ti	100%	75%	50%	25%
35	0.736	0.548	0.403	0.298
40	0.838	0.633	0.473	0.356
45	0.957	0.733	0.555	0.425
50	1.096	0.849	0.652	0.505
55	1.259	0.984	0.765	0.599
60	1.447	1.142	0.897	0.709
65	1.665	1.325	1.050	0.838
70	1.918	1.537	1.228	0.987
75	2.209	1.781	1.434	1.160
80	2.543	2.063	1.672	1.360
85	2.926	2.386	1.945	1.590
90	3.364	2.757	2.259	1.854
95	3.864	3.180	2.618	2.155
100	4.432	3.662	3.027	2.499
105	5.077	4.210	3.492	2.889
110	5.807	4.831	4.019	3.331
115	6.630	5.532	4.614	3.831
120	7.557	6.322	5.285	4.393
125	8.596	7.208	6.038	5.025

3 Pump Generator Output [MWe]

Ti	100%	75%	50%	25%
35	1175.6	863.5	557.7	216.8
40	1177.6	866.3	558.8	216.1
45	1179.5	868.9	559.5	215.0
50	1181.1	871.0	559.7	213.6
55	1182.2	872.5	559.2	211.8
60	1182.6	873.1	557.9	209.6
65	1182.1	872.8	555.6	206.9
70	1180.4	871.3	552.3	203.8
75	1177.2	868.3	547.9	200.2
80	1172.3	863.8	542.3	196.2
85	1165.5	857.5	535.4	191.7
90	1156.6	849.3	527.2	186.8
95	1145.3	839.1	517.7	181.5
100	1131.6	826.9	506.9	175.9
105	1115.4	812.6	494.9	170.0
110	1096.5	796.4	481.9	164.0
115	1075.2	778.2	467.9	157.7
120	1051.4	758.3	453.0	151.3
125	1025.3	736.8	437.4	144.9

Condenser Heat Reject [MBTU/hr]

Ti	100%	75%	50%	25%
35	7521.7	5702.2	3861.6	2138.1
40	7515.0	5692.7	3857.8	2140.7
45	7508.6	5684.2	3855.5	2144.3
50	7503.3	5677.2	3854.9	2149.0
55	7499.6	5672.2	3856.6	2155.0
60	7498.3	5669.9	3861.0	2162.4
65	7500.0	5671.0	3868.5	2171.4
70	7505.8	5676.2	3879.5	2181.8
75	7516.4	5686.1	3894.2	2193.8
80	7532.7	5701.3	3913.1	2207.3

85	7555.4	5722.4	3936.2	2222.2
90	7585.3	5749.8	3963.6	2238.6
95	7623.0	5783.8	3995.4	2256.2
100	7668.8	5824.6	4031.4	2275.0
105	7723.2	5872.2	4071.4	2294.6
110	7786.2	5926.6	4115.0	2315.0
115	7857.6	5987.3	4161.9	2335.9
120	7937.2	6053.9	4211.6	2357.1
125	8024.4	6125.7	4263.6	2378.5

Condenser Rise [øF]

Ti	100%	75%	50%	25%
35	20.51	15.55	10.53	5.83
40	20.49	15.52	10.52	5.84
45	20.47	15.50	10.51	5.85
50	20.46	15.48	10.51	5.86
55	20.45	15.47	10.52	5.88
60	20.44	15.46	10.53	5.90
65	20.45	15.46	10.55	5.92
70	20.46	15.48	10.58	5.95
75	20.49	15.50	10.62	5.98
80	20.54	15.54	10.67	6.02
85	20.60	15.60	10.73	6.06
90	20.68	15.68	10.81	6.10
95	20.78	15.77	10.89	6.15
100	20.91	15.88	10.99	6.20
105	21.06	16.01	11.10	6.26
110	21.23	16.16	11.22	6.31
115	21.42	16.32	11.35	6.37
120	21.64	16.51	11.48	6.43
125	21.88	16.70	11.62	6.49

Appendix F. Cooling Pond Performance

The following functions implement Langhaar's model for cooling ponds. These may be used to model the thermal performance of large ponds.

```
/*********************************************************
***********************
  POND-GEOMETRY
*********************************************************
**********************/

double PondArea(double Elevation)                  /* pond
area in acres from elevation */
  {                                                /* in feet
above mean sea level */
  if(Elevation<=583.5)
    return(0);
  return(252.305*pow(Elevation-583.5,0.945033));
  }

double PondVolume(double Elevation)          /* pond volume
in acre-feet from elevation */
  {                                                /* in feet
above mean sea level */
  if(Elevation<=583.5)
    return(0);
  return(252.305*pow(Elevation-
583.5,1.945033)/1.945033);
  }

/*********************************************************
***********************
  LANGHAAR'S POND CALCULATIONS
*********************************************************
**********************/

double yRH(double RH)
  {
  return((81.3295615276*RH-
359.043847242)*RH+566.189533239);
  }

double yTX(double Tdb)
  {
  return((-0.00364208459429*Tdb-
0.938531836914)*Tdb+563.661164074);
  }

double yTY(double Tdb)
  {
```

```
  return((((-
0.000011002057396*Tdb+0.00312849291125)*Tdb-
0.363391922646)*Tdb+24.0624509974)*Tdb-268.381574621);
  }

double fTE(double y)
  {
  return(((-7.64388403125E-8*y+1.767707201209E-
4)*y+0.02270881300489)*y+21.22088945338);
  }

double Teq1(double RH,double Tdb)
  {
  double x1,x2,x3,y1,y2,y3;
  x1=116.;
  y1=yRH(RH);
  x2=yTX(Tdb);
  y2=yTY(Tdb);
  x3=659.;
  y3=y1+(y2-y1)*(x3-x1)/(x2-x1);
  return(fTE(y3-1.9073576336026));
  }

double yTE(double Te)
  {
  return(((2.445370284702E-4*Te-
.0728401209826)*Te+12.12846891621)*Te-77.3036975845);
  }

double gTE(double y)
  {
  return(((-1.514543864717E-7*y+3.173255575958E-4)*y-
0.02245149639398)*y+21.17697337907);
  }

double yWS(double WS)
  {
  return((0.113846153846*WS-7.4)*WS+485.153846154);
  }

double fQS(double y)
  {
  return(exp((3.28103282914E-7*y-
0.005219646991425)*y+7.629928327939));
  }

double yQS(double Qs)
  {
  double z;
```

```
  z=log(Qs);
  return((2.732299407884*z-
232.2979147965)*z+1614.086511135);
  }

double fWs(double X)
  {
  return((-0.06*X+4.34)*X+700.8);
  }

double Teq2(double Teq1,double WS,double Qa)
  {
  double Qs,x1,x2,x3,y1,y2,y3,y4,y5;
  x1=85.;
  x2=336.;
  x3=519.;
  y1=yTE(Teq1);
  y2=yWS(WS);
  y3=y1+(y2-y1)*(x3-x1)/(x2-x1);
  Qs=fQS(y3+1.70916334657522);
  y4=yQS(Qs+Qa);
  y5=y4+(y2-y4)*(x1-x3)/(x2-x3);
  return(gTE(y5+0.874316939823416));
  }

double yE(double Te)
  {
  return((((3.06219996606E-6*Te-
.00119039629931)*Te+0.169554749432)*Te-
4.71740208339)*Te-.703035058119);
  }

double Ey(double y)
  {
  return(0.174879844254*y+41.0017947314);
  }

double gWS(double WS)
  {
  if(WS<9.5)
    return((-
1.13687436159*WS+43.2414708887)*WS+23.2145045965);
  if(WS<13.)
    return(((-0.572918896334*WS-
31.496518252)*WS+310.040267259)/(1.-0.118033856491*WS));
  if(WS<16.)
    return((-
0.366371018114*WS+15.7249219238)*WS+224.693316677);
  if(WS<25.)
```

```
    return((((-0.00208301885643*WS+0.189635219107)*WS-
6.29716977817)*WS+93.6137012375)*WS-143.871291359);
  if(WS<35.)
    return((((0.000265697376144*WS-
0.0337665004286)*WS+1.58346148339)*WS-
28.5970835609)*WS+559.210936962);
  if(WS<40.)
    return((((0.00080000001463*WS-
0.102666668718)*WS+4.88000010623)*WS-
97.93333574)*WS+1100.00002011);
  return((((1.6690358468E-6*WS-
0.00859920943636)*WS+1.24997058155)*WS-
53.9210443316)*WS+1171.17549936);
  }

double gQ(double y)
  {
  return(exp(-0.00514002689607*y+1.18818451182));
  }

double fQ(double Te,double WS)
  {
  double x1,x2,x3,y1,y2,y3;
  x1=60.;
  x2=225.;
  x3=452.;
  y1=yE(Te);
  y3=gWS(WS);
  y2=y1+(y3-y1)*(x2-x1)/(x3-x1);
  return(gQ(y2+5.10204089863748E-2));
  }

double yDTi(double dTi)
  {
  double z;
  z=log(dTi);
  return(((3.50768670016*z-16.5785425742)*z-
132.330053787)*z+664.85176737);
  }

double yP(double P)
  {
  return(-1.14828363384*P+312.083287101);
  }

double fDTo(double y)
  {
```

```
  return(exp((((4.89112665649E-11*y-4.85669595205E-
8)*y+0.000017963796707)*y+0.00360401184359)*y+0.47386888
4976));
  }

double dTe(double dTi,double P)
  {
  double x1,x2,x3,y1,y2,y3;
  x1=68.;
  x2=300.;
  x3=529.;
  y1=yDTi(dTi);
  y2=yP(P);
  y3=y1+(y2-y1)*(x3-x1)/(x2-x1);
  return(fDTo(y3+0.943965516979461));
  }

void PondTables()
  {
  double ep,P,Qa=100.,RH,Tdb,Teq,WS;

  printf("\n  Pond Geometry\n");
  printf(" Elev Area Volum\n");
  printf(">MSL  acre ac-ft\n");
  for(ep=583.5;ep<590.1;ep+=0.5)
    printf("%5.1lf %4.0lf
%5.0lf\n",ep,PondArea(ep),PondVolume(ep));

  printf("\nLanghaar's Pond Performance\n");

  printf("\nEquilibrium Temperature\n");
  printf("Tdb   0%%  25%%  50%%  75%% 100%%\n");
  for(Tdb=20.;Tdb<101.;Tdb+=10.)
    {
    printf("%3.0lf",Tdb);
    for(RH=0.;RH<1.01;RH+=0.25)
      printf(" %4.1lf",Teq1(RH,Tdb));
    printf("\n");
    }

  printf("\nAdjusted Equilibrium Temperature\n");
  printf("Teq     0     5     10     15     20 mph\n");
  for(Teq=30.;Teq<141.;Teq+=10.)
    {
    printf("%3.0lf",Teq);
    for(WS=0.;WS<20.1;WS+=5.)
      printf(" %5.1lf",Teq2(Teq,WS,Qa));
    printf("\n");
    }
```

```
  printf("\n       Scaling Factor, Q\n");
  printf("Teq    0     5     10     15     20 mph\n");
  for(Teq=30.;Teq<141.;Teq+=10.)
    {
    printf("%3.0lf",Teq);
    for(WS=0.;WS<20.1;WS+=5.)
      printf(" %5.3lf",fQ(Teq,WS));
    printf("\n");
    }

  printf("\n               Approach\n");
  printf("Teq  P=50   100   150   200   250\n");
  for(Teq=30.;Teq<141.;Teq+=10.)
    {
    printf("%3.0lf",Teq);
    for(P=50.;P<251.;P+=50.)
      printf(" %5.2lf",dTe(Teq,P));
    printf("\n");
    }

}/****************************************************
```

The function PondTables() lists the calculated performance for a range of input variables and also illustrates how to call the functions. The output is as follows:

```
  Pond Geometry
 Elev Area Volum
>MSL  acre ac-ft
583.5    0     0
584.0  131    34
584.5  252   130
585.0  370   285
585.5  486   499
586.0  600   771
586.5  713  1099
587.0  824  1483
587.5  935  1923
588.0 1045  2418
588.5 1155  2968
589.0 1264  3573
589.5 1372  4232
590.0 1480  4945

Langhaar's Pond Performance

Equilibrium Temperature
Tdb    0%   25%   50%   75% 100%
 20 20.6 20.9 21.3 21.8 22.2
 30 24.6 26.1 27.5 28.9 30.1
 40 31.1 33.5 35.9 38.0 39.9
```

50	38.0	41.4	44.5	47.4	49.9
60	44.8	49.1	53.1	56.7	59.9
70	51.5	56.8	61.7	66.1	69.9
80	58.1	64.5	70.3	75.6	80.1
90	64.0	71.6	78.5	84.6	89.9
100	68.1	76.9	84.9	92.0	98.1

Adjusted Equilibrium Temperature

Teq	0	5	10	15	20 mph
30	66.1	58.3	52.8	49.2	46.7
40	72.3	65.2	60.2	56.9	54.7
50	79.0	72.5	68.0	65.0	63.0
60	85.8	80.0	76.0	73.3	71.6
70	92.7	87.5	84.0	81.6	80.1
80	99.7	95.1	92.0	89.9	88.6
90	106.7	102.7	100.0	98.2	97.0
100	114.0	110.4	108.1	106.5	105.5
110	121.5	118.5	116.4	115.1	114.2
120	129.5	126.9	125.2	124.1	123.4
130	138.1	135.9	134.5	133.6	133.0
140	147.3	145.6	144.5	143.8	143.3

Scaling Factor, Q

Teq	0	5	10	15	20 mph
30	3.304	2.201	1.647	1.533	1.483
40	2.996	1.996	1.494	1.390	1.345
50	2.628	1.750	1.310	1.219	1.180
60	2.254	1.501	1.123	1.046	1.012
70	1.908	1.271	0.951	0.885	0.857
80	1.606	1.070	0.801	0.745	0.721
90	1.351	0.900	0.673	0.627	0.606
100	1.138	0.758	0.567	0.528	0.511
110	0.960	0.640	0.479	0.445	0.431
120	0.811	0.540	0.404	0.376	0.364
130	0.682	0.454	0.340	0.316	0.306
140	0.568	0.379	0.283	0.264	0.255

Approach

Teq	P=50	100	150	200	250
30	13.13	6.20	2.98	1.64	1.44
40	16.48	7.72	3.67	1.90	1.39
50	19.47	8.99	4.27	2.14	1.41
60	22.13	10.07	4.78	2.35	1.45
70	24.50	11.00	5.21	2.54	1.50
80	26.61	11.79	5.58	2.70	1.55
90	28.47	12.46	5.90	2.84	1.59
100	30.12	13.04	6.16	2.96	1.63
110	31.56	13.53	6.39	3.06	1.67
120	32.81	13.95	6.58	3.15	1.70
130	33.89	14.31	6.74	3.22	1.73
140	34.82	14.61	6.88	3.28	1.75

Appendix G. Power Plant Thermal Simulation

The functions presented in Appendix E and F are combined to simulate the thermal response of this nuclear power plant. The code is provided below and in the on-line archive (filename nuke.c). The results were shown in Chapter 10.

```
double Required(/* required pond area [ft²/gpm] */
  double Tdb,    /* dry-bulb temperature [øF] */
  double RH ,    /* relative humidity (0 to 1) */
  double GHI,    /* solar input [BTU/hr/ft²] */
  double mph,    /* wind speed [mph] */
  double Thw,    /* inlet (hot) water temp [øF] */
  double Tcw)    /* exit (cold) water temp [øF] */
  {
  double dT1,dT2,P,Q,Teq;
  if(Tcw<=Teq1(RH,Tdb))
    return(FLT_MAX);
  if(Thw<=Tcw)
    return(0.);
  Teq=Teq2(Teq1(RH,Tdb),mph,GHI);
  if(Teq<33.)
    Teq=33.;
  dT1=Thw-Teq;
  dT2=Tcw-Teq;
  if(dT2<=0.)
    return(FLT_MAX);
  P=fPDT(dT1,dT2);
  Q=fQ(Teq,mph);
  return(P*Q);
  }

double bp[3]; /* backpressures (zones 1, 2, 3) [in.Hg]
*/
double Tc[3]; /* condenser temperatures (zones 1, 2, 3)
[øF] */
double Tpond; /* pond temperature [øF] */
double mwe  ; /* generator output [MWe] */

void Thermal( /* thermal operating point */
  double Tdb,    /* dry-bulb temperature [øF] */
  double RH ,    /* relative humidity (0 to 1) */
  double GHI,    /* solar input [BTU/hr/ft²] */
  double mph,    /* wind speed [mph] */
  double elev,   /* pond elevation [ft above MSL] */
  int pumps ,    /* number of CCW pumps on */
  double mwt)    /* reactor heat input [MWt] */
  {
  double Apond; /* pond area [acres] */
  double Di   ; /* condenser inside tube diameter [in]
*/
```

```
  double gpm  ; /* condenser flow [gpm] */
  int iter;
  double Tp1,Tp2;
  Apond=PondArea(elev); /* pond surface area */
  Di=Dtube-2.*Wall[gage-12]; /* inside diameter */
  gpm=CondenserFlow(pumps,Ntubes,Di,Ltube); /* condenser
flow */
  Tp1=max(32.,Teq1(RH,Tdb)); /* upper and lower bound */
  Tp2=150.; /* on pond temperature */
  for(iter=0;iter<32;iter++) /* use a bisection search
to determine pond temp */
    {
    Tpond=(Tp1+Tp2)/2.;

mwe=Electrical(mwt,gpm,Ltube,Ntubes,cleanliness,material
,Dtube,Di,Tpond,bp,Tc);

if(Required(Tdb,RH,GHI,mph,Tc[2],Tpond)*gpm/43560.>Apond
)
      Tp1=Tpond;
    else
      Tp2=Tpond;
    }
  }

void Simulation(char*fname)
  {
  char bufr[128];
  int pumps=3;
  double elev=590.,GHI,mph,mwt=3425.,RH,Tdb;
  FILE*fp;
  fp=fopen(fname,"rt");
  while(fgets(bufr,sizeof(bufr),fp))
    {
    if(sscanf(bufr,"%lf%*[ ,\t]%lf%*[ ,\t]%lf%*[ ,\t]
%lf%*[ ,\t]",&Tdb,&RH,&mph,&GHI)==4)
      {
      if(Tdb<10.)
        Tdb=10.;
      Thermal(Tdb,RH,GHI,mph,elev,pumps,mwt);
      printf("%lG,%lG,%lG,%lG\n",Tpond,Tc[2],bp[2],mwe);
      }
    }
  fclose(fp);
  }

int main(int argc,char**argv,char**envp)
  {
  if(argc<2)
```

```
  {
  PlantTables();
  PondTables();
  }
else
  Simulation(argv[1]);
return(0);
}
```

Appendix H. Nitrogen Supersaturation Models

The following code and data were used in the nitrogen supersaturation model for Jennings Randolph Project.

```
#include <stdio.h>
#include <stddef.h>
#include <stdlib.h>
#include <float.h>
#include <math.h>

/* USACE WES Nitrogen SuperSaturation Model */

/* constants and empirical coefficients */

double Cs=   100; /* saturation concentration [%] */
double Cu=   100; /* upstream concentration [%] */
double g =   9.8; /* gravitational acceleration
[m/sec^2] */
double Pa=101325; /* atmospheric pressure [N/m^2] */
double Sc=   0.6; /* Schmidt number */
double Vr=  0.25; /* bubble rise velocity [m/sec] */

double alpha =  4.69; /* mass transfer parameter */
double beta  =   2.2; /* effective depth parameter */
double eta   =  0.66; /* power on Reynolds number */
double Ks    =  1.77; /* surface transfer parameter */
double lambda=  0.19; /* air layer thickness [m] */
double mu    =.00078; /* dynamic viscosity of water
[kg/m/sec] */
double rho   =  1000; /* density of water [kg/m^3] */
double sigma =0.0720; /* surface tension [N/m] */

#define gamma (rho*g)  /* specific weight of water
[N/m^3] */
#define nu    (mu/rho) /* kinematic viscosity of water
[m^2/sec] */

double Supersaturation(
  double Hr, /* river depth [m] */
  double Hs, /* stilling basin depth [m] */
  double Ht, /* total head [m] */
  double Ls, /* stilling basin length [m] */
  double Dp, /* maximum plunge depth [m] */
  double q)  /* flow per unit width [m^2/sec] */
  {
  double bH; /* dimensionless bH=beta*Hb/L */
  double Ce; /* effective saturation concentration [%]
*/
  double Ci; /* infinite flow concentration [%] */
```

```
  double Db; /* effective bubble depth [m] */
  double Dr; /* effective river depth [m] */
  double Ds; /* effective stilling basin depth [m] */
  double Dj; /* depth of plunging jet [m] */
  double Fi; /* volumetric air concentration */
  double Hb; /* bubble half-life distance [m] */
  double Kb; /* bubble mass transfer coefficient */
  double Rb; /* bubble rise velocity Reynolds number */
  double Re; /* Reynolds number */
  double rK; /* increase in mass transfer coefficient */
  double Vj; /* jet velocity [m/sec] */
  double We; /* Weber number */

  Db=2*Hs/3;
  Dr=Hr/2;
  Re=q/nu;
  Hb=log(2)*q/Vr;
  bH=beta*Hb/Ls;
  if(bH>1)
    Ds=Dr+(Db-Dr)*exp(1-bH);
  else
    Ds=Db;
  if(Ds>Dp)
    {
    rK=Ds/Dp;
    Ds=Dp;
    }
  else
    rK=1;
  Rb=2*Ds*Vr/nu;
  Vj=.15*pow(q/sqrt(g*Ht)/Ht,.23)*sqrt(g*Ht);
  Fi=Vj*lambda/(Vj*lambda+q);
  Dj=q/Vj;
  We=rho*q*q/sigma/Dj;
  Kb=alpha*Fi*sqrt(1-Fi)/pow(1-
pow(Fi,5/3.),.25)*pow(We,.6)*pow(Re,eta)/sqrt(Sc)/Rb;
  Ce=Cs*(1+Ds*gamma/Pa);
  Ci=(rK*Kb*Ce+rK*Ks*Cs)/(rK*Kb+rK*Ks);
  return(Ci-(Ci-Cu)*exp(-rK*Kb-rK*Ks));
  }

double Weir(double Cu,double e)
  {
  return(Cu+e*(Cs-Cu));
  }

double Downstream(double Cu,double X,double Q)
  {
  double L=15079+2.99*Q;
```

```
  return(Cu+(Cs-Cu)*(1-exp(-X/L)));
  }

#define FEET_PER_METER 3.2808399

/* site specific data and computed parameters */

double Dp=12.0; /* maximum plunge depth [m] */
double Hr= 2.8; /* river depth [m] */
double Hs=18.0; /* stilling basin depth [m] */
double Ht= 5.0; /* total head [m] */
double Ls=95.4; /* stilling basin length [m] */
double Ws=19.1; /* width of spillway [m] (62'8") */

double Model(double e,double X,double q)
  {

return(Downstream(Weir(Supersaturation(Hr,Hs,Ht,Ls,Dp,q)
,e),X,q*pow(FEET_PER_METER,3)*Ws));
  }

struct{
  double Ce;
  double Cs;
  double Ks;
  double Kb;
  double Tb;
  }mT;

void MassTransfer(double t,double*C,double*dC)
  {
  dC[0]=mT.Kb*(mT.Ce-C[0])+mT.Ks*(mT.Cs-C[0]);
  if(t<mT.Tb)
    dC[0]=mT.Kb*(mT.Ce-C[0]);
  else
    dC[0]=mT.Ks*(mT.Cs-C[0]);
  }

void RungeKutta4(             /* one step of 4th order
Runge-Kutta integration */
  void dydx(double,double*,double*),  /* function
returning dY/dX */
  double*x,                   /* independent variable
*/
  double dx,                  /* step in X */
  double*y,                   /* dependent variable
array */
  double*dy,                  /* step in Y array */
```

```
  int n)                    /* number of dependent
variables */
  {
  int i;
  double*v,*w1,*w2,*w3,*w4;

  v=calloc(5*n,sizeof(double));
  w1=v+n;
  w2=w1+n;
  w3=w2+n;
  w4=w3+n;

  dydx(x[0],y,w1);

  for(i=0;i<n;i++)
    v[i]=y[i]+w1[i]*dx/2;
  dydx(x[0]+dx/2,v,w2);

  for(i=0;i<n;i++)
    v[i]=y[i]+w2[i]*dx/2;
  dydx(x[0]+dx/2,v,w3);

  for(i=0;i<n;i++)
    v[i]=y[i]+w3[i]*dx;
  dydx(x[0]+dx,v,w4);

  for(i=0;i<n;i++)
    {
    dy[i]=(w1[i]+2*w2[i]+2*w3[i]+w4[i])/6;
    y[i]=y[i]+dy[i]*dx;
    }
  x[0]+=dx;

  free(v);
  }

double Cx=260,Qx=45;

void Results(
  char*site,
  double Hs   , /* stilling basin depth [m] */
  double Hr   , /* river depth [m] */
  double Ht   , /* total head [m] */
  double Ls   , /* stilling basin length [m] */
  double*data, /* data pairs (q,Kb) */
  int n,    /* number of data pairs */
  int set)
  {
  int i;
```

```
  double q;
  double bX; /* dimensionless bX=beta*X2/L */
  double Db; /* effective bubble depth [m] */
  double De; /* effective depth [m] */
  double Dr; /* effective river depth [m] */
  double X2; /* bubble half-life distance [m] */

  printf("\nmodel parameters for %s\n",site);
  printf("  Hs=%9.3lG stilling basin depth [m]\n",Hs);
  printf("  Hr=%9.3lG river depth [m]\n",Hr);
  printf("  Ht=%9.3lG total head [m]\n",Ht);
  printf("  Ls=%9.3lG stilling basin length [m]\n",Ls);
  printf("\ncomputed parameters for %s\n",site);

  Db=2*Hs/3;
  printf("  Db=%9.3lG effective bubble depth [m]\n",Db);

  Dr=Hr/2;
  printf("  Dr=%9.3lG effective river depth [m]\n",Dr);

  printf("\n  q  Cse\n");
  printf("%s measured\n",site);
  for(i=0;i<n;i++)
    {
    q=data[2*i];
    printf("%4.1lf %3.0lf\n",q,data[2*i+1]);
    printf("%i %lG %lG\n",set,q,data[2*i+1]);
    }

  printf("%s computed\n",site);
  for(q=Qx;q>FLT_EPSILON;q/=1.1)
    {
    X2=log(2)*q/Vr;
    bX=beta*X2/Ls;
    if(bX>1)
      De=Dr+(Db-Dr)*exp(1-bX);
    else
      De=Db;
    printf("0 %lG %lG\n",q,100*(1+De*gamma/Pa));
    }
  }

double IceHarbor[]={
   3.2,115,
   3.4,121,
   6.3,127,
   6.5,126,
   9.7,130,
   9.7,136,
```

```
  12.7,136,
  12.8,140,
  15.9,138,
  16.2,136,
  18.9,135,
  19.3,134,
  22.0,132,
  25.5,130,
  28.6,128,
  31.7,127,
  31.3,124,
  34.1,124,
  37.2,122};

double TheDalles[]={
   3.3,114,
   3.2,116,
   5.8,115,
   5.5,117,
   8.3,115,
   7.3,117,
   8.4,118,
   9.6,118,
  10.9,117,
  11.2,119,
  12.6,118,
  13.2,118,
  13.6,117,
  16.7,116,
  19.0,114,
  19.3,115,
  21.4,113};

double LittleGoose[]={
   3.5,105,
   7.1,113,
   7.0,116,
  10.7,122,
  10.6,128,
  14.1,128,
  14.2,131,
  17.8,130,
  17.7,132,
  17.7,134,
  21.3,135,
  24.8,138,
  28.2,139,
  31.4,140,
  34.6,139,
```

```
  35.2,137,
  37.9,137,
  41.3,138};

double JenningsRandolph[]={
  3.71,117,
  3.71,116,
  3.71,117,
  3.71,118,
  4.45,117,
  5.19,120,
  5.93,119,
  6.67,122,
  6.67,121,
  6.67,124,
  6.67,118,
  7.41,118,
  7.41,125,
  7.41,126,
  7.86,121,
  8.15,121,
  8.15,122,
  8.60,121,
  8.89,126,
  9.34,117,
 -10.1,128, /* extrapolation */
 -14.1,132,
 -16.0,134,
 -18.1-12,135,
 -20.1-12,132,
 -22.0-12,130,
 -24.1-12,128,
 -26.0-12,127,
 -28.0-12,126,
 -29.9-12,124,
 -32.0-12,123,
 -34.1-12,122,
 -36.0-12,121,
 -38.1-12,120,
 -40.2-12,119,
 -42.0-12,118,
 -44.0-12,118,
 -46.1-12,117,
 -48.1-12,117,
 -50.2-12,116,
 -52.1-12,116,
 -54.1-12,116,
 -56.1-12,115,
 -58.1-12,115,
```

```
 -60.0-12,115};

double erfc(double X)                              /*
approximation for the error function */
  {                                                      /*
from Abramowitz & Stegun */
  static double C0= 0.327591100;                   /* Handbook
of Mathematical Functions */
  static double C1= 0.254829592;
  static double C2=-0.284496736;
  static double C3= 1.421413741;
  static double C4=-1.453152027;
  static double C5= 1.061405429;
  double Q,T,Y;

  Y=fabs(X);
  T=1/(1+C0*Y);
  Q=((((C5*T+C4)*T+C3)*T+C2)*T+C1)*T;

  if(X<0)
    return(2-Q/exp(X*X));
  else
    return(Q/exp(X*X));
  }

#define erf(x) (1-erfc(x))

double SuperSat(double Qx,double Cu,double Cx,double Q)
  {
  static double A=-4.33407,a=-
0.00326463,B=4.84528,b=0.706924,c=0.945739;
  return(Cu+(Cx-Cu)*(A*exp(-
a*Q/Qx)*erf(b*Q/Qx)+B*erf(c*Q/Qx)));
  }

#undef gamma
#undef nu

int main(int argc,char**argv,char**envp)
  {
  double gamma,nu;

  printf("USACE WES Nitrogen SuperSaturation Model\n");

  printf("\nconstants and empirical coefficients\n");
  printf("  alpha =%9.31G mass transfer
parameter\n",alpha);
  printf("  beta  =%9.31G effective depth
parameter\n",beta);
```

```
  printf("  Cs    =%9.3lG saturation concentration
[%%]\n",Cs);
  printf("  eta   =%9.3lG power on Reynolds
number\n",eta);
  printf("  g     =%9.3lG gravitational acceleration
[m/sec^2]\n",g);
  printf("  Ks    =%9.3lG surface transfer
parameter\n",Ks);
  printf("  lambda=%9.3lG air layer thickness
[m]\n",lambda);
  printf("  mu    =%9.3lG dynamic viscosity of water
[kg/m/sec]\n",mu);
  printf("  Pa    =%9.0lf atmospheric pressure
[N/m^2]\n",Pa);
  printf("  rho   =%9.0lf density of water
[kg/m^3]\n",rho);
  printf("  Sc    =%9.3lG Schmidt number\n",Sc);
  printf("  sigma =%9.3lG surface tension
[N/m]\n",sigma);
  printf("  Vr    =%9.3lG bubble rise velocity
[m/sec]\n",Vr);

  printf("\ncomputed parameters\n");
  gamma=rho*g;
  printf("  gamma =%9.0lf specific weight of water
[N/m^3]\n",gamma);

  nu=mu/rho;
  printf("  nu    =%9.3lG kinematic viscosity of water
[m^2/sec]\n",nu);

  Results("Ice Harbor"  ,12.2, 4.3,29.0,59.4,IceHarbor
,sizeof(IceHarbor)/sizeof(double)/2,1);
  Results("The Dales"   , 6.8, 2.8,24.7,42.7,TheDalles
,sizeof(TheDalles)/sizeof(double)/2,2);
  Results("Little
Goose",21.3,11.3,29.6,88.4,LittleGoose,sizeof(LittleGoos
e)/sizeof(double)/2,3);

  return(0);
  }
```

also by D. James Benton

3D Articulation: Using OpenGL, ISBN-9798596362480, Amazon, 2021 (book 3 in the 3D series).

3D Models in Motion Using OpenGL, ISBN-9798652987701, Amazon, 2020 (book 2 in the 3D series.

3D Rendering in Windows: How to display three-dimensional objects in Windows with and without OpenGL, ISBN-9781520339610, Amazon, 2016 (book 1 in the 3D series).

A Synergy of Short Stories: The whole may be greater than the sum of the parts, ISBN-9781520340319, Amazon, 2016.

Azeotropes: Behavior and Application, ISBN-9798609748997, Amazon, 2020.

bat-Elohim: Book 3 in the Little Star Trilogy, ISBN-9781686148682, Amazon, 2019.

Boilers: Performance and Testing, ISBN: 9798789062517, Amazon 2021.

Combined 3D Rendering Series: 3D Rendering in Windows®, 3D Models in Motion, and 3D Articulation, ISBN-9798484417032, Amazon, 2021.

Complex Variables: Practical Applications, ISBN-9781794250437, Amazon, 2019.

Compression & Encryption: Algorithms & Software, ISBN-9781081008826, Amazon, 2019.

Computational Fluid Dynamics: an Overview of Methods, ISBN-9781672393775, Amazon, 2019.

Computer Simulation of Power Systems: Programming Strategies and Practical Examples, ISBN-9781696218184, Amazon, 2019.

Contaminant Transport: A Numerical Approach, ISBN-9798461733216, Amazon, 2021.

CPUnleashed! Tapping Processor Speed, ISBN-9798421420361, Amazon, 2022.

Curve-Fitting: The Science and Art of Approximation, ISBN-9781520339542, Amazon, 2016.

Death by Tie: It was the best of ties. It was the worst of ties. It's what got him killed., ISBN-9798398745931, Amazon, 2023.

Differential Equations: Numerical Methods for Solving, ISBN-9781983004162, Amazon, 2018.

Equations of State: A Graphical Comparison, ISBN-9798843139520, Amazon, 2022.

Forecasting: Extrapolation and Projection, ISBN-9798394019494, Amazon 2023.

Heat Engines: Thermodynamics, Cycles, & Performance Curves, ISBN-9798486886836, Amazon, 2021.

Heat Exchangers: Performance Prediction & Evaluation, ISBN-9781973589327, Amazon, 2017.

Heat Recovery Steam Generators: Thermal Design and Testing, ISBN-9781691029365, Amazon, 2019.

Heat Transfer: Heat Exchangers, Heat Recovery Steam Generators, & Cooling Towers, ISBN-9798487417831, Amazon, 2021.
Heat Transfer Examples: Practical Problems Solved, ISBN-9798390610763, Amazon, 2023.
The Kick-Start Murders: Visualize revenge, ISBN-9798759083375, Amazon, 2021.
Jamie2: Innocence is easily lost and cannot be restored, ISBN-9781520339375, Amazon, 2016-18.
Kyle Cooper Mysteries: Kick Start, Monte Carlo, and Waterfront Murders, ISBN-9798829365943, Amazon, 2022.
The Last Seraph: Sequel to Little Star, ISBN-9781726802253, Amazon, 2018.
Little Star: God doesn't do things the way we expect Him to. He's better than that! ISBN-9781520338903, Amazon, 2015-17.
Living Math: Seeing mathematics in every day life (and appreciating it more too), ISBN-9781520336992, Amazon, 2016.
Lost Cause: If only history could be changed..., ISBN-9781521173770, Amazon, 2017.
Mass Transfer: Diffusion & Convection, ISBN-9798702403106, Amazon, 2021.
Mill Town Destiny: The Hand of Providence brought them together to rescue the mill, the town, and each other, ISBN-9781520864679, Amazon, 2017.
Monte Carlo Murders: Who Killed Who and Why, ISBN-9798829341848, Amazon, 2022.
Monte Carlo Simulation: The Art of Random Process Characterization, ISBN-9781980577874, Amazon, 2018.
Nonlinear Equations: Numerical Methods for Solving, ISBN-9781717767318, Amazon, 2018.
Numerical Calculus: Differentiation and Integration, ISBN-9781980680901, Amazon, 2018.
Numerical Methods: Nonlinear Equations, Numerical Calculus, & Differential Equations, ISBN-9798486246845, Amazon, 2021.
Orthogonal Functions: The Many Uses of, ISBN-9781719876162, Amazon, 2018.
Overwhelming Evidence: A Pilgrimage, ISBN-9798515642211, Amazon, 2021.
Particle Tracking: Computational Strategies and Diverse Examples, ISBN-9781692512651, Amazon, 2019.
Plumes: Delineation & Transport, ISBN-9781702292771, Amazon, 2019.
Power Plant Performance Curves: for Testing and Dispatch, ISBN-9798640192698, Amazon, 2020.
Practical Linear Algebra: Principles & Software, ISBN-9798860910584, Amazon, 2023.
Props, Fans, & Pumps: Design & Performance, ISBN-9798645391195, Amazon, 2020.
Remediation: Contaminant Transport, Particle Tracking, & Plumes, ISBN-9798485651190, Amazon, 2021.

ROFL: Rolling on the Floor Laughing, ISBN-9781973300007, Amazon, 2017.
Seminole Rain: You don't choose destiny. It chooses you, ISBN-9798668502196, Amazon, 2020.
Septillionth: 1 in 10^{24}, ISBN-9798410762472, Amazon, 2022.
Software Development: Targeted Applications, ISBN-9798850653989, Amazon, 2023.
Software Recipes: Proven Tools, ISBN-9798815229556, Amazon, 2022.
Steam 2020: to 150 GPa and 6000 K, ISBN-9798634643830, Amazon, 2020.
Thermochemical Reactions: Numerical Solutions, ISBN-9781073417872, Amazon, 2019.
Thermodynamic and Transport Properties of Fluids, ISBN-9781092120845, Amazon, 2019.
Thermodynamic Cycles: Effective Modeling Strategies for Software Development, ISBN-9781070934372, Amazon, 2019.
Thermodynamics - Theory & Practice: The science of energy and power, ISBN-9781520339795, Amazon, 2016.
Version-Independent Programming: Code Development Guidelines for the Windows® Operating System, ISBN-9781520339146, Amazon, 2016.
The Waterfront Murders: As you sow, so shall you reap, ISBN-9798611314500, Amazon, 2020.
Weather Data: Where To Get It and How To Process It, ISBN-9798868037894, Amazon, 2023.

www.ingramcontent.com/pod-product-compliance
Lightning Source LLC
Chambersburg PA
CBHW020920180526
45163CB00007B/2818
* 9 7 8 1 5 2 0 9 1 3 3 4 6 *